The New York Botanical Garden

The New York Botanical Garden

An Illustrated Chronicle of Plants and People

Ogden Tanner
and Adele Auchincloss

WALKER AND COMPANY
NEW YORK

First published in the United States of America in 1991
by Walker Publishing Company, Inc.

Published simultaneously in Canada by Thomas Allen & Son
Canada, Limited, Markham, Ontario

Library of Congress Cataloging-in-Publication Data

Tanner, Ogden.
The New York Botanical Garden : an illustrated chronicle of plants and people / by Ogden Tanner, Adele Auchincloss.
Includes bibliographical references and index.
ISBN 0-8027-1141-3
1. New York Botanical Garden. I. Auchincloss, Adele. II. Title.
QK73.U62N497 1991
580'.74'47471—dc20 90-43704
CIP

Printed in Hong Kong

2 4 6 8 10 9 7 5 3 1

Text design by Georg Brewer

Acknowledgements

The authors are grateful to the many individuals at The New York Botanical Garden who helped with this book. Special thanks go to John Reed, vice-president for education and director of the library, who coordinated their efforts, and to Allen Rokach, former staff photographer, who provided most of the pictures.

Thanks also to others who furnished photographs, including Marcia Stevens and Muriel Weinerman of the Garden's photographic staff, and Michael Balick, Joseph Beitel, Alan Berkowitz, Alan Bolten, Brian Boom, William Buck, Andrew Cott, Douglas Daly, Michael Doolittle, Ann Marie Gaynor, Carol Gracie, Roy Halling, Noel Holmgren, Maria Lebrón, James Luteyn, Celia Maguire, Scott Mori, Christine Padoch, Robert Perron, Ghillean Prance, Michael Ruggiero, Ted Spiegel, Barbara Thiers and Wayt Thomas.

Contents

A View from Kew

A Foreword by Ghillean Prance

—DIRECTOR, ROYAL BOTANIC GARDENS
KEW, RICHMOND, SURREY, ENGLAND

WE AT THE ROYAL BOTANIC GARDENS, KEW, ARE PROUD THAT OUR INSTITUTION helped inspire Nathaniel Lord Britton and his wife to start The New York Botanical Garden a century ago. The pride is well justified: in science, education and horticulture, NYBG now ranks among the foremost in the world.

The connection means even more to me personally, and in a way completes the circle. The New York Botanical Garden was my own training ground for 25 years before taking over as director of Kew.

I think those years prepared me well: the work, and the mission, of the two gardens are much the same. In fact, when a problem comes up at Kew, my colleagues have frequently been surprised to hear me say that it is old hat, because I experienced the identical situation whilst at NYBG.

When I left New York in 1988 it was with considerable sadness, particularly at leaving behind so many good friends. I served NYBG under three presidents. When I first arrived, I was made to feel most welcome by William Steere, who was then in the process of rebuilding many aspects of the Garden, including its scientific work. My introduction was through the person who later became Steere's successor, Howard Irwin, whom I met to join a 1963 expedition in Suriname. The training I received from Howard, a true professional in the field, stood me in good stead for the next 24 years, much of which was spent exploring the Amazonian region on behalf of NYBG.

What impressed me about the Garden's field work was its high standard, not to mention its productivity, both of which continue with the fine group of botanists who work there today. I am grateful, too, to Jim Hester, under whose presidency I learned much about administration, including many ideas that I have used in my present job.

There are others I remember fondly. Imagine my surprise, as a young Scotsman, at discovering that a Lancashire native with a broad north-country English accent was in charge of horticulture at the Garden. I enjoyed Tom Everett, with his wicked sense of humour and his repertoire of risqué stories. More importantly, he showed me that horticulture was a vital part of NYBG's mission. A tour of the T. H. Everett Rock Garden, with its creator, was an education that no young botanist could ever forget. In fact, it was also excellent training for Kew!

Today, botanical gardens no longer have the luxury of simply carrying out basic research and inventory, although the importance of these activities has certainly not diminished. We must also address the environmental destruction that is a worldwide feature of this century. I was happy to be part of the development of the Garden's Institute of Economic Botany and Institute of Ecosystem Studies, which are helping to focus NYBG's role and show its readiness to face the challenge of the times. The Garden is an ideal place for the institutes to do their work: they are supported by a dynamic programme in systematic botany, and by the largest herbarium, and the best botanical library, in the United States. (I have many fond memories of the latter, too, especially of such marvelous characters, and infinite sources of wisdom, as Lothian Lynas and the late Miss Elizabeth Hall.)

Where NYBG has gone way beyond Kew, I think, is in adult and children's education. I was privileged to observe the ways in which Arnold Gussin and John Reed built up a great programme from a modest start. The knowledge that I gained from these two gifted educators will certainly be put to use here over the next few years.

And so the Royal Botanic Gardens, and its director, congratulate The New York Botanical Garden on its first century, and wish it every success for its second.

Global Missions

An Introduction

As The New York Botanical Garden celebrates its 100th anniversary, it faces an era of critical change, one that is already shaping its character for the years ahead. The crisis, which is rapidly becoming a central issue of the 1990s, is the deteriorating state of the world's environment, along with an alarming extinction of plant and animal species that threatens to erase the biological diversity on which the planet's future depends. As Harvard biologist E. O. Wilson reminds us, permitting that loss of diversity is the one folly our descendants will be least likely to forgive.

It is a crisis of our own making. The Earth's human population, which has doubled since the early 1950s to more than 5.3 billion, is projected by the United Nations to reach 6.25 billion by the end of this century, and, without more vigorous efforts to control growth, could double again by the end of the next century, reaching the staggering total of 14 billion before it stabilizes. With more than 90 percent of the increase concentrated in the developing nations, the UN observes, the numbers of poor and hungry will escalate, as will damage to the environment from the pressures of people on land, forests and water supplies, as well as the risk of global climate change.

It is this relentless multiplication of a single species, the National Science Foundation (NSF) observes, that threatens to raise the extinction of other life forms to at least 1,000 times the normal rate. What we are already beginning to witness, says NSF, is the most catastrophic loss of species since the natural eradication of the dinosaurs and their environment 65 million years ago.

But where in this sobering picture, its friends might ask, does the Garden fit in? To answer that, one must go back a bit.

Botanical gardens are among man's oldest and most treasured institutions, with spiritual roots in the Eden of Adam and Eve. For much of their long history they were focused almost entirely on useful plants—a focus that is returning with new urgency today. Gardens of medicinal species existed in China thousands of years ago, and in other ancient cultures as scattered as those of the Aztecs and Egyptians. In the fourth century B.C., Aristotle established in Athens what was probably the first garden designed for the scientific study of plants, turning it over in his will to his pupil and friend Theophrastus, widely regarded as the "father of botany." The tradition was carried on through medieval times in monasteries, which became experimental grounds and sources of herbs for all manner of practical ends.

The first true botanical gardens of modern times, starting with those at Padua and Pisa in the mid-sixteenth century, were administered by universities, which compiled their collections like encyclopedias of living plants, celebrating God's creations that supplied man's food, drink and pleasure and His divine "physick" that could cure men's ills.

By the eighteenth and nineteenth centuries, physic gardens abounded, many under the auspices of medical schools, whose teachers were as much botanists as they were physicians. Gradually, however, it was seen that such gardens had other roles to play, including the discovery and display of new ornamental species from near and far, as well as informing the public in the knowledge and culture of plants. It was into this dawning era of threefold responsibility—science, horticulture and education—that The New York Botanical Garden was born.

During its first century, the Garden has taken on still broader dimensions, as have many of its counterparts around the world. It has grown into a major museum—a modern encyclopedia of plants, with more than 25,000 kinds of living plants and a herbarium of over 5 million preserved specimens. It has also blossomed into a horticultural university, an international research center, a scientific command post that has sent nearly 900 expeditions to all corners of the globe. The keystone of its work continues to be systematic botany: the discovery, classification and comparative study of plants with the hope of understanding the evolutionary history, and marvelous diversity, of the natural world in which we live.

In recent decades, that goal has become crucial. Only now are we beginning to realize how much we depend on plants, our silent green companions on this earth. Amid increasingly man-made surroundings, they provide us with a psychological link to nature whose importance to human health is only now being clinically proved. Even more basically, plants have been the great producers and supporters of life on earth since the first organisms capable of photosynthesis appeared billions of years ago. Plants give us the oxygen we breathe, the foods we eat, the fuels we burn, the timber we build with, the paper we write on, the fibers we wear, the medicines that soothe our aches and save our lives. Yet of perhaps a million plant species only a quarter have been identified and only a minute percent have been investigated in any detail.

The most alarming aspect of the crisis is that time is running out. In the tropics, where two-thirds of the world's plant species are thought to exist, deforestation is proceeding at a rate estimated at 50 to 100 acres a minute, the equivalent of a football field every second. Much of the diversity of animal life in these regions, not to mention the well-being of humans, is threatened by the destruction of vegetation that provides both habitat and food. Of 6,000 kinds of legumes in the tropics, scientists believe as many as 4,000 are in danger of extinction. These and other plants that we may need for our survival, including new sources of treatment for diseases like cancer and AIDS, could be gone forever—before we even know what they are, or what vital roles they might have played. Yet because of a shortage of scientists trained in systematic botany, and a low funding priority in both the public and private sectors, no truly comprehensive survey of tropical plants yet exists.

The New York Botanical Garden is one of the places where such scientists work. Through its programs in systematics, and its Institute of Economic Botany, NYBG is actively engaged in surveying the flora of the tropics, and in helping to develop from it new sources of food, medicine and energy. Through its Institute of Ecosystem Studies, and its educational programs, the Garden is also striving to protect and improve the quality of the environment—to achieve, and to impart to others, a better understanding of the complex natural webs on which all life depends.

Over its first 100 years, The New York Botanical Garden has compiled a distinguished record. But as a new force in conservation and environmental education, it has only begun.

The New York Botanical Garden

Glories of the Garden

When you look a little closer, a new world opens up.

Allen Rokach goes out into the garden when other people don't, and he sees things they often miss. As NYBG's former staff photographer for 15 years, he has made thousands of stunning visual images—patterns of beauty seen through an artist's eye. Some of his finest, shown on the pages that follow, reveal new ways to look at gardens, and at nature, whether you have a camera along or not.

To Rokach, the first secret is timing. "Most visitors come to the Garden in the middle of a bright, sunny day, when the light is strong and flat and the colors are washed out," he says. "I like to go out in the early morning, when the light is soft and fresh, the birds and animals are more active and I have the place almost to myself. Another good time is in the slanting, golden light of late afternoon, or later, just at dusk."

Like other skilled observers, Rokach does much of his image-hunting when the sky is overcast and the true richness of colors and textures come to the fore. One of his favorite times is on a misty day or right after a rainstorm when everything seems more alive because it is dripping wet. He roams the Garden at different seasons, even in the dead of winter, when he can discover subtle patterns and colors that appear at no other time of year. He often makes a game of trying to identify bare-branched species just by the different ways they hold the snow.

A final secret is patience, along with attention to detail. "A lot of visitors to the Garden see only big splashes of color when the daffodils or azaleas are in bloom. They ooh and ah, then rush on to the next thing," he says. "You have to be willing to spend a little time, to walk around, to search for the essence. Don't just look at the

◄

The Barbara Foster Vietor Memorial Walk glows with as many as 120 varieties of tulips in spring. In the background, a glimpse of the Museum Building's dome. (Allen Rokach)

flowers; go up and look *inside* the flowers, at the petals and other graceful parts. Often you'll see insects going about their work, which also reminds you what flowers are all about.

"When you look a little closer, you see things in a way that you haven't before," Rokach concludes. "And when that happens, a whole new world begins to open up." (All the photographs in this chapter were taken by Allen Rokach.)

The Conservatory dome, seen across the herb garden fence.

WINTER

An upright European hornbeam (Carpinus betulus *'Fastigiata') casts long shadows on the snow.*

The sinuous form of a threadleaf Japanese maple (Acer palmatum dissectum), *etched against gray skies.*

A winter scene on the Bronx River.

Dried seedheads of swamp rose-mallow (Hibiscus palustris) *in the Swale.*

The Hester Bridge, arching over the Bronx River gorge.

Snow patterns on a saucer magnolia (Magnolia × soulangiana), *with a European beech* (Fagus sylvatica) *behind.*

▶
Flowering dogwoods (Cornus florida), *hybrid rhododendrons and azaleas bloom on Azalea Way.*

SPRING

In Cherry Valley, flowering cherries (Prunus) *put on a spring display.*

Eastern redbuds (Cercis canadensis) *light up the woods.*

'Queen of Sheba', a dazzling lily-flowered tulip on Vietor Walk.

◄

Chinese, or garden, peonies (Paeonia lactiflora *'Doreen'*).

European pasqueflower (Anemone pulsatilla), *a Rock Garden favorite.*

▶

A spider flower cultivar, Cleome hasslerana *'Rose Queen', with* Zinnia *'Gold Sun'.*

SUMMER

In summer, the Vietor Walk becomes a showcase for mixed annuals, including marigolds, zinnias and spider flowers.

In the Irwin Perennial Garden, the bright plumes of Astilbe × arendsii *'Peach Blossom' rise above hosta and yucca leaves.*

Yellow California poppies (Eschscholzia californica) *and a purple hybrid sage* (Salvia × superba *'Ost Friesland'*).

►

Ruby Swiss chard lends color to the Loeb Vegetable Garden.

Sedum *'Autumn Joy' and mealycup sage* (Salvia farinacea *'Victoria'*).

▶

A threadleaf Japanese maple (Acer palmatum dissectum) *against the sky.*

FALL

The Bronx River in fall.

Patterns of meadow grasses.

▶

Autumn in the Irwin Garden, with plumes of zebra grass (Miscanthus sinensis *'Zebrinus'*).

A cutleaf European white birch (Betula alba *'Dalecarlia'*) *and the vivid leaves of burning bush* (Euonymus alatus).

Willows and oaks in the Swale on a fall afternoon.

Turkey-tail mushrooms (Trametes versicolor) *on a log.*

◀ *Flowerheads of* Astilbe taquetii *'Superba'.*

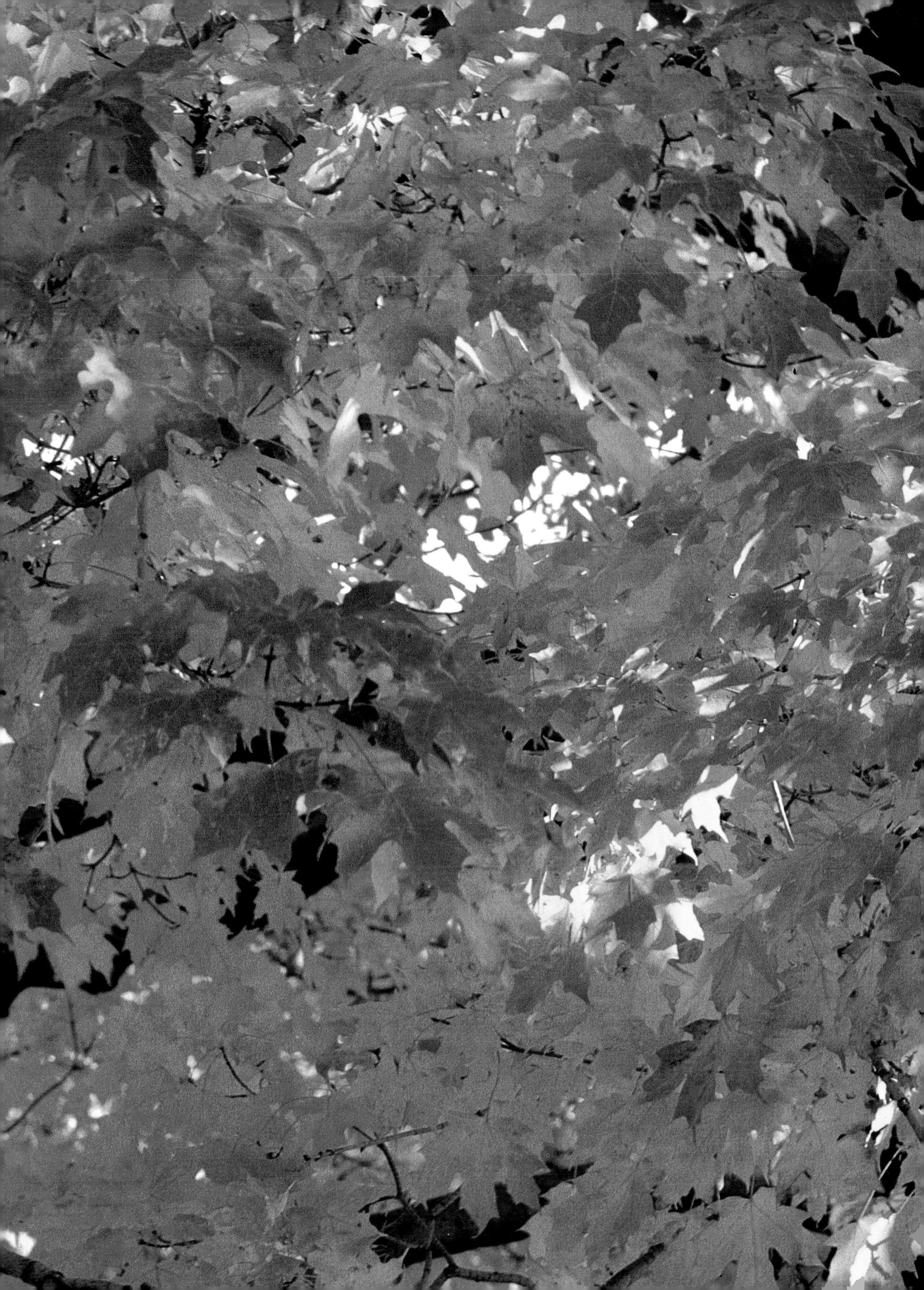

Sugar maple (Acer saccharum).

An American Kew

"Oh, why can't we have a garden just like that!"

New York in the 1880s was a booming, brawling place, a city of political bosses who ruled by bribery and smoked big cigars, of immigrants who sailed past the new Statue of Liberty with stars in their eyes and wound up sleeping in the streets.

It was also a city where business was good, and the rich were very rich. People with names like Vanderbilt and Astor built grand palazzos on Fifth Avenue, read about their latest extravagances in the papers and often spent their summers abroad, where they ordered their clothes in Paris and London and their marble statues in Rome. Captivated by the Old World, these new patrons of culture determined to equal its glories at home. Civic leaders had already built a Philharmonic Hall and an American Museum of Natural History, and had just opened a new Metropolitan Opera and Metropolitan Museum of Art. Only a major botanical garden, a zoo and a central public library were missing from the list.

The notion of a botanical garden had come up before. In 1877, a consortium was authorized to issue $25 shares for a public display garden on the fringes of Central Park, a dream that evaporated when the needed 14,000 shares could not be sold. In 1883, others petitioned for new greenswards to serve the city's burgeoning population, which was rapidly moving north. As a result of their efforts, 3,757 acres of largely open land were earmarked under a new parks act for Pelham Bay, Van Cortlandt and Bronx parks.

The idea of using one of these parks for a botanical garden started at a meeting of the Torrey Botanical Club, a group of ladies and gentlemen who met periodically in

◄

A nature lover contemplates the quiet beauty of the hemlock grove along the Bronx River. (NYBG Archive)

In the early days, open fields were hayed with scythes and horse-drawn mowers, providing fodder for the Garden's livestock. (NYBG Archive)

Manhattan to discuss the finer points of plants (and who had named the organization after John Torrey, the eminent New York botanist who had originated the idea decades before by holding informal seminars in his home).

On October 24, 1888, it was noted in the minutes, one of the members, Elizabeth Britton, "gave a description of the Botanical Establishment at Kew." She and her husband had recently returned from a belated honeymoon in England, where they had spent many happy hours at the Royal Botanic Gardens outside London, considered the finest in the world for both its displays and its scientific expertise.

Her husband later confided to a friend that on one of their outings, Elizabeth had turned to him and exclaimed, "Oh, why can't we have a garden just like that!"

A committee of other members appointed to look into the matter agreed that New York indeed deserved, as they put it, "a public botanic garden of the highest class." They cautioned that its primary purposes should be scientific and educational, but admitted that it could also be "a place of agreeable resort for the public at large."

Galvanized by their newfound mission, Torrey devotees published appeals in the newspapers and placed posters around town. The ladies of the club, led by Elizabeth Britton, mounted a heroic campaign of luncheons, teas and dinners in the homes of New York's wealthiest patrons, at which various speakers extolled the virtues of a botanical garden for the "betterment of man." Among the speakers was her husband, then 30 years old and about to be promoted to professor of botany at Columbia College (later to become Columbia University). Though slight of stature, Nathaniel Lord Britton's forceful speaking style was evidently impressive. So, perhaps, was his fortuitous family name.

Among the plan's advocates were two federal judges, Addison Brown and Charles Daly, who had lifelong interests in botany and horticulture. In 1891 an act they had drawn up for the state legislature was passed, authorizing the city's Department of Parks to set aside 250 acres in Bronx Park for a New York Botanical Garden (at about the same time, additional acreage was earmarked for the New York Zoological Society, which was to build its Bronx Zoo a few years later in the southern part of the park). The charter marked the official beginnings of NYBG:

> *. . . for the purpose of establishing and maintaining a botanical garden and museum and arboretum therein, for the collection and culture of plants, flowers, shrubs and trees, the advancement of botanical science and knowledge, and the prosecution of original researches therein and in kindred subjects, for affording instruction in the same, for the prosecution and exhibition of ornamental and decorative horticulture and gardening, and for the entertainment, recreation and instruction of the people.*

A charter was one thing; money was another. The city agreed to underwrite a bond issue for necessary buildings and improvements up to the sum of $500,000 but stipulated that the Garden's backers raise an additional $250,000 in private funds to demonstrate their commitment to the public the Garden was going to serve.

To most New Yorkers of a century ago, a quarter of a million dollars was a daunting sum. The Garden's promoters, however, had thoughtfully cultivated, and elected as the principal officers of their new corporation, three of the most powerful men in America—Cornelius Vanderbilt, Andrew Carnegie and J. Pierpont Morgan (see box). This august triumvirate attended the first official board meeting in the office of Seth Low, president of Columbia College. Keeping a close eye on the proceedings, as the corporation's new secretary, was Nathaniel Lord Britton himself.

New Yorkers in their Sunday best visit the new Conservatory soon after its opening at the turn of the century. (NYBG Archive)

Cornelius Vanderbilt (NYBG Archive)

Andrew Carnegie (NYBG Archive)

J. Pierpont Morgan (The Museum of the City of New York)

Blue Ribbon Backers

WHEN NEW YORKERS HEARD THE NEWS THAT THEY WERE ABOUT TO GET A new botanical garden, reactions ranged from "How nice!" to "What's *that?*" But if anyone questioned the seriousness of the project, they had only to look at the *New York Tribune* of June 19, 1895. The list of the Garden's backers read like a distillation of *Who's Who in America,* with a generous dash of the *Social Register* thrown in.

Serving as president was Cornelius Vanderbilt II, grandson of the commodore and director of the family's vast railroad empire. Vanderbilt, whose family were friends of the Brittons, was quite possibly the richest man in America at the time.

Vice-president was Andrew Carnegie, the famed Scottish immigrant and philanthropist, who had started at $1.20 a week in a bobbin factory and amassed a fortune in iron and steel. According to an estimate by J. Pierpont Morgan, Carnegie was the richest man in the *world.*

Holding the purse strings as treasurer was Morgan himself, known to the business community as the most powerful banker in America and the nation's economic czar. A model for the epithet "robber baron," Morgan was celebrated for being both honest and ruthless, a reputation that proved of considerable help in his business dealings. While not cowing lesser mortals, Morgan took great joy in his family, his art collections, his yachts and his civic largesse.

In addition to lending sage business advice and fund-raising clout, this trio donated $25,000 apiece to help get the Garden on its feet. Equal amounts were contributed by John D. Rockefeller, Darius Ogden Mills, Judge Addison Brown and Columbia University. Rounding out the list were other prominent New York names—Gould, Dodge, Sloan, Schermerhorn—as well as Arnold, Constable and Tiffany, two of the city's finest stores.

Together, the founding patrons provided not only money. They started a tradition of public service that has attracted civic leaders to the Garden ever since. Descendants of early trustees, in fact, have continued to serve as members of NYBG's past and present boards.

THE SUBSCRIBERS

Columbia College	$25,000	Oswald Ottendorfer	5,000
J. Pierpont Morgan	25,000	Samuel Sloan	5,000
Andrew Carnegie	25,000	George J. Gould	5,000
Cornelius Vanderbilt	25,000	Helen M. Gould	5,000
John D. Rockefeller	25,000	John S. Kennedy	5,000
D. O. Mills	25,000	William Rockefeller	5,000
Hon. Addison Brown	25,000	Arnold, Constable & Co.	5,000
William E. Dodge	10,000	Morris K. Jesup	2,500
James A. Scrymser	10,000	Mrs. Melissa P. Dodge	1,000
Wm. C. Schermerhorn	10,000	Tiffany & Co.	1,000
Hon. Charles P. Daly	5,000	Hugh N. Camp	500
			$250,000

With other luminaries joining in, the money was eventually raised. The effort was measurably boosted when some 50 "women of wealth" met at the residence of Mrs. Louis Fitzgerald and listened to Bishop Henry Potter stress the "elevating influence of the ministry of flowers in life." There were also dark moments when fund-raising almost foundered in the stock market panic of 1893.

For the site of the Garden, the directors selected the northern section of Bronx Park, which at 719 acres was the smallest but most central of the city's three new parks. It also had good public transportation, being on the New York Central's Harlem railroad line (a fact that may not have been lost on Cornelius Vanderbilt, the railroad's president).

Best of all, the site embraced a magnificently varied landscape of rolling fields, water features, rock outcroppings and native trees, including a 40-acre grove of towering hemlocks along the Bronx River that had been decreed forever wild. Before it had come into the city's hands in 1884 as a result of the new parks act, the land had been part of the 661-acre estate of Pierre Lorillard, a leading tobacco merchant. Lorillard had dammed the river and erected a mill for the manufacture of snuff, which he scented with petals from his large rose garden nearby, and had also built a 45-room mansion, stables and a gatekeeper's house. Happily, Lorillard had respected the natural beauty of the land, and his sturdy stone buildings could be put to use.

There were a few problems. A swampy area deemed malarial would have to be dealt with. Some directors were concerned that an abundance of saloons on nearby roads would make the Garden difficult to police, and debated whether to erect a fence. The board also began to realize that even $250,000 would not be adequate to the task ahead, and thought it wise to cultivate contributing memberships from the "public at large." By the end of the following year 440 members had been enlisted at a minimum of $10 apiece.

In May 1896, during a fact-finding mission to the site, the directors met aboard one of Vanderbilt's trains en route to the Bronx and appointed The New York Botanical Garden's first director-in-chief—Nathaniel Lord Britton.

A master plan for the new garden was drawn up by a special commission, consisting of Britton, landscape engineer John Brinley, landscape gardener Samuel Henshaw, Professor Lucien Underwood of the Garden's Board of Scientific Directors, architect Robert Gibson and Lincoln Pierson of Lord & Burnham, the nation's leading greenhouse firm, which was to build the conservatory. The plan showed various roads and paths, as well as specific outdoor growing areas: a Fruticetum for fruiting plants, a Viticetum for vines, a Pinetum, a Deciduous Arboretum, an Herbaceous Grounds for flowers, an Economic Garden for useful plants, a Bog Garden and a Rockery, as well as substantial dwellings for the director and the two chief gardeners. On a slight rise convenient to the existing railroad station, the scheme displayed a monumental Museum Building, designed by Gibson and containing more than 90,000 square feet of

The Museum Building, with Corinthian columns and a copper-sheathed dome, nears completion in 1901. (NYBG Archive)

space. Set back farther into the grounds (in front of what are now the Rock and Native Plant Gardens) would stand a glittering conservatory of iron and glass for tropical plants labeled "First Horticultural House."

The plan touched off heated arguments between its backers, who pictured the Garden as a great scientific institution, and those who thought the land should remain an oasis of pristine beauty as a public park. The debate raged in the pages of the *New York Sun,* the *New York Tribune* and *Harper's Weekly,* whose editors agreed with the "parkies" that the directors were "not free to treat the land given to them . . . as if they had bought and paid for it." (One reporter, apparently tiring of the hue and cry, ventured that the Bronx air of the 1890s was too smoky for the health of plants in any case.) The city's superintendent of parks, Samuel Parsons, wanted the hemlock forest excluded from Garden jurisdiction, but relented when Cornelius Vanderbilt personally pledged to preserve it as a public treasure. The officers did make a few concessions: the elegant staff houses were deleted, and the Conservatory was relocated on a less conspicuous rise to the south, where there was more space. Then they went ahead with their plans.

On the last day of 1897, ground was broken for the Museum Building, the first in the nation devoted solely to botany, with 19-foot-high exhibition halls, a herbarium of preserved specimens and administrative offices on the upper floors, all faced with leafy Corinthian columns and surmounted by a copper dome. The ornate Beaux Arts design, awarded in an architectural competition to Gibson, took some three years to build at a cost of more than $250,000, reaching completion in 1901. Even more impressive, and quite different from any other structure yet seen in New York, was the $180,000 Conservatory for the display of tender and tropical plants, whose airy, crystalline pavilions were first opened to the public in 1900 (see Chapter 4).

Garden and city officials, in full regalia, break ground for the Conservatory building in January, 1899. (NYBG Archive)

Meanwhile, under the direction of Britton and his head gardener Samuel Henshaw, whom he had hired at the princely sum of $50 per month, workmen were busy all over the grounds, clearing undergrowth, laying paths, planting flowers, seeding lawns, putting labels on trees, establishing a nursery as a source of additional plants. The troublesome marshy area to the north was dredged, and Lorillard's dam was lowered to provide better drainage, converting the area into a pair of pretty ponds that were christened Twin Lakes. In front of the Museum Building, Charles Tefft's sculpture *Fountain of Life* was installed. Twenty-four tulip trees were planted in two

(NYBG Archive)

Growing Plants —and Skyscrapers

NYBG WAS NOT THE FIRST BOTANICAL GARDEN IN NEW YORK. IN FACT, ONE operated for a while in the city almost two centuries ago, though it is less famous for its pioneering efforts than what eventually happened to its site.

Dr. David Hosack, who had trained under eminent botanists and physicians in Edinburgh and London, devoted his life to the cultivation of plants, particularly those of medicinal value. As a practicing physician and Professor of Materia Medica at Columbia's College of Physicians and Surgeons, he had tried to persuade the state of New York to establish a garden that would serve both the medical profession and the public at large. When his efforts proved unsuccessful, in 1801 Dr. Hosack purchased 20 acres of open land in central Manhattan—at that time, a three-mile carriage ride from downtown—and turned them into his own garden, complete with a handsome central greenhouse built of stone and glass. Here he grew plants of medicinal and botanical interest, many obtained from scientist friends. He also grew vegetables and fruits, which he gave them in return. Among his students were Asa Gray and John Torrey, who were to become illustrious botanists. Leading horticulturists of the day often came to his garden to exchange plants and ideas.

Hosack, who named his establishment the Elgin Garden after his father's birthplace in Scotland, accumulated some 2,000 specimens of exotic and native plants, but found that the operation was too expensive for his limited resources. When annual costs exceeded $100,000 in 1810, he was forced to sell the garden—for $4,807.36—to the state of New York, which showed little interest in maintaining it and finally turned it over to Columbia College in 1814. Columbia, still struggling to establish itself, had neither the interest nor the money to keep the garden going, and tried renting it to commercial horticulturists. When this proved unprofitable, the college leased some of the land in parcels for business and housing, retaining title to most of the lots.

With the passage of time, what had been a failed enterprise became the most celebrated success story in the history of Manhattan real estate. In 1928 John D. Rockefeller, Jr., paid Columbia the unheard-of sum of $100 million for 200 parcels covering less than three blocks. Today the former Elgin Garden is the site of Rockefeller Center, the first planned concentration of skyscrapers in the United States, among other things noted for its use of gardens on its rooftops and along its central promenade. Even before Rockefeller's purchase, the investment was making possible the expansion of Columbia into a major university—which became a major backer, and continuing supporter, of The New York Botanical Garden that was finally built in the Bronx.

stately rows leading down to a low, walled court (the latter unfortunately obliterated when NYBG's blocky, utilitarian Laboratory Building was erected in 1956).

The Garden's earliest plantings included formal Victorian beds, with flowers displayed in straight rows and patterns near the entrance, and more informal settings inspired by wilder portions of the site. *The New York Times* reported in the summer of 1900 that the shrub collection was coming along nicely, with the addition of 17 different varieties of barberry bushes, raspberries and blackberries, old varieties of roses, sumacs, spireas and hydrangeas.

"The North Meadow is a delight to people who like nature as she is," the paper noted. "Along the river there are fine willow trees already growing [and a] quantity of joe-pye weed. . . . There is the border that begins at the station and runs along the garden, following the railroad track—a broad border of old-fashioned flowers with a background of green. There are two miles [sic] of this blossoming boundary, and last week it was beautiful with rose mallows and phlox, with sunflowers, too, and the jimson weed, a beautiful, ladylike plant."

The *Times* also gave readers a fine glimpse of the new director-in-chief at work: "The best way to see the grounds, if one is fortunate enough to be able to do so, is with Dr. Britton and his pony. That pony takes him over the grounds usually about three times a day, sometimes riding, but driving if he has company. There is nowhere that the pony, after long practice, cannot go. . . . It has been said that he can climb trees, if the doctor cares to make investigations as high as that."

The new Garden soon became a favorite destination for New Yorkers, as many as 15,000 of whom traveled to the Bronx by train, trolley or carriage on Sundays to marvel at its horticultural delights. It was reported that residents of Manhattan's crowded Lower East Side tenements found the place especially rewarding. Many spent sweltering summer weekends camped out in the coolness of the hemlock grove along the Bronx River, where lines of laundry could often be seen hung out to dry.

Whatever he thought of such intrusions, Britton was not deterred from his tasks. In 1905 a series of public lectures was initiated in the basement of the Museum Building, and courses in nature study were started for children of the Bronx and Manhattan public school systems. In 1906 the Garden began to use docents to conduct public tours of its horticultural displays. Heavy use of the Garden led the directors to start building a boundary fence of iron and stone along major city roads, though the grounds would not be fully enclosed until 1940.

Aspiring gardeners watch a NYBG instructor demonstrate the proper use of a spade. (NYBG Archive)

Well dressed, well mannered school children embark on a guided tour of the Garden's myriad wonders. (NYBG Archive)

Although women had played a key role in the Garden's beginnings, recognition of the fact was slow. In 1913 it was proposed that three ladies be added to the Board of Managers; the idea was vetoed by the all-male members. A year later a women's advisory council was created, and some of its members eventually wound up on the board. Described by one of its later leaders as "ladies who arrived at the Garden in limousines ready to give advice and large sums of money," the group conducted fund-raising benefits and obtained new dues-paying members (renamed the NYBG Council, it remains important today, taking on special projects as well as helping to support the Children's Garden).

By 1915, when the city added 140 acres to the Garden's holdings, as many as 50,000 people were visiting the grounds on pleasant weekends. In 1917, Daniel and Murry Guggenheim, prominent New York businessmen and members of NYBG's board, donated $50,000 each for the construction of additional conservatories on the eastern edge of the Garden's new land. Opened to the public two years later was a complex of greenhouses fronted by a 170-foot-long display pavilion, which had space for lectures among plants from warm temperate regions of South Africa, South America, southern Japan and the southern United States. Nearby, another new greenhouse accommodated the Garden's growing orchid collections. The Guggenheim conservatories were in use until the late 1930s, when the city, in the person of the redoubtable parks commissioner, Robert Moses, took much of the land back to build the Bronx River Parkway, and the buildings were torn down.

Meanwhile, the Garden was pursuing its scientific goals. William Dodge, a leading Wall Street financier and member of the board, provided funds for NYBG's first plant-hunting expedition, in which Per Axel Rydberg, a young Swedish botanist, was sent to Montana in 1897 to collect plants of America's montane West. The following year Cornelius Vanderbilt underwrote an exploration of Puerto Rico by a Mr. and Mrs. A. A. Heller, who returned with a collection of some 8,000 specimens. Before long the

▶

On a plant-hunting expedition to the West Indies, Director Britton (right) examines a towering species of the cactus family. (NYBG Archive)

Brittons themselves were journeying to the West Indies; the director's personal interest in the region was to lead to nearly a hundred NYBG expeditions there in the early half of the century.

Under the watchful eye of the Brittons, the Garden continued to grow. When Nathaniel Lord Britton finally stepped down in 1929, he could take pride in some of America's finest living plant collections, a herbarium of 1.7 million specimens, a library of 43,500 bound volumes and widely respected programs in science and public education—an accomplishment that some of Britton's peers ranked second only to the much older botanical gardens at Kew and Berlin.

Thanks to his single-minded ambition, and a little help from well-placed friends, in less than four decades the New York Botanical Garden had become one of the leading institutions of its kind in the world.

As part of the war effort, volunteers in uniform tended NYBG's Victory Garden, demonstrating how patriotic citizens could grow their own food. (NYBG Archive)

His Garden

(NYBG Archive)

NATHANIEL LORD Britton—the name was always used in full—may not have been a certified nobleman, but, as one of his associates drily noted, he could sometimes act the part. Born to an old Staten Island family of landowners in 1859, he spent boyhood days roaming the fields and shores near his home, collecting shells, stones and especially plants. He graduated from Columbia College with an Engineer of Mines degree, joined the staff as an instructor in geology and botany, and later received his Ph.D. in botany after completing a thesis on the flora of New Jersey.

Relatively unknown when he was appointed NYBG's first director, Britton used his engineer's training in helping to design the Garden's buildings and grounds. His true love, however, was botany, particularly scientific research, in which he excelled; he came to specialize in the flora of the West Indies, where he and his wife Elizabeth made 30 or more expeditions, usually at their own expense. Britton's publications included the first studies of the plants of Bermuda and the Bahamas, as well as extensive surveys of Cuba and Puerto Rico. He and Elizabeth liked the latter so much they often spent winter vacations there, staying at the Condado Vanderbilt Hotel and hiring a large Packard car and chauffeur. Around the island—where a mountain was later named after Britton honoring his work—he was sometimes referred to as "the millionaire botanist."

A prolific author, during his lifetime Britton produced 880 scientific papers and books, including such major works as *North American Trees* (1908) and *The Cactaceae* (1920, with J. N. Rose—see Chapter 4). Probably his greatest contribution, with Addison Brown, was *The Illustrated Flora of the Northern United States, Canada and the British Possessions,* a three-volume work published in 1896 that laid the groundwork for ongoing studies of North American flora by later NYBG scientists. Britton had little patience with fellow botanists who withheld their findings, sometimes for years, until they could perfect them. Among Garden colleagues he was well known for his exhortation: "Get it into print!"

Nathaniel Lord Britton was no giant physically: he stood about five feet three inches tall and weighed less than 110 pounds (it was a common joke that his beard, which grew longer as he grew older, accounted for much of his weight). Generally mild-spoken, when it came to NYBG business he could be a martinet—some used the term "czar"—insisting that the Garden was entitled to the full days, and nights, if need be, of its staff. He allowed employees no outside professional activities without his permission, which he seldom gave. Britton considered electric lights to be an unnecessary luxury—gas lamps were good enough, as long as they were used sparingly—and fought a losing battle against installing electricity in NYBG's offices and laboratories.

The director gave generously of his inheritance to Garden projects and expected others to do the same. An associate recalls that Britton wrote personal letters to each member of the board whenever money was needed for a new expedition or a special collection of books, stating that he was giving such and such and requesting that the recipient contribute at least half that amount. And, the associate noted, he usually got it.

After his official retirement, the Brittons continued their travels and studies until Elizabeth died in 1934. It was a shock from which her husband never recovered. Confined to bed by a stroke, Nathaniel Lord Britton died four months later at the age of 75.

(NYBG Archive)

. . . and Hers

THERE IS LITTLE doubt that Britton's drive was shared by Elizabeth Gertrude Knight, whom he married in 1885. They were to have no children, but with their passion for plants, and particularly for their beloved Garden, they led a full and generally happy life.

A native New Yorker, Elizabeth spent early summers on her grandfather's sugar plantation in Cuba, where she learned to speak fluent Spanish. She graduated at the age of 17 from what is now Manhattan's Hunter College and stayed on for a while as a tutor of natural science. To advance a growing interest in mosses, she joined the Torrey Botanical Club, where, in all probability, she met her husband-to-be.

Mrs. Britton not only suggested the idea of a New York Botanical Garden, she worked hard to make it a reality over the years, showing up almost every day at the Museum Building at 10 A.M. She served as honorary curator of mosses, in which she had become a leading authority, helping to develop an extensive bryological collection for the Garden's herbarium. Among employees she was known as a lady, but one who moved with a quick, nervous walk—and had a loud, clear reprimand for anyone whom she thought deserved it. ("She was a rather domineering sort of woman," one staff member recalled. "She made her presence known.")

The Garden aside, Elizabeth Britton's greatest contribution to horticulture was the Wildflower Preservation Society of America, which she helped to found in 1902. Among 346 publications that bear her name, the most charming are her essays on such wildflowers as the azalea, the jack-in-the-pulpit and the columbine. Determined to save native species from exploitation, she urged a national boycott on the collecting of American holly as a Christmas decoration, proposing that nurseries propagate plants instead. In honor of her pioneering role, a special fund was established for a new wildflower garden at NYBG, begun in the 1930s. It is now known as the Native Plant Garden, and a plaque there bears her name.

A 250-Acre Showcase

A BIRD'S-EYE VIEW OF THE NEW YORK BOTANICAL GARDEN SHOWS MAJOR gardens, plantings and buildings, some of them pictured around the border. Most are within easy walking distance of the main gate and visitor information center. Numbers on the map are identified below.

1. Metro North Train Station
2. Main Garden Gate
P. Parking
3. Tulip Tree Mall and Café
4. Museum Building
   Visitor Information Center
   Shop in the Garden
   Herbarium
   Library
5. Auditorium
6. Watson Education Building
7. Beauty Bush Pavilion
8. Compass Garden
9. Demonstration Gardens
10. Enid A. Haupt Conservatory
11. Perennial Garden
12. Herb Garden
13. Rock Garden
    Native Plant Garden
14. NYBG Forest
15. Swale Gazebo
16. Daffodil Hill
17. Crab Apples
18. Conifers
19. Snuff Mill Restaurant
20. Propagation Facility
21. Rose Garden
22. Service Buildings
23. Cherry Valley
24. Daylily Garden
25. Children's Garden
26. Waterfall
27. Magnolias
28. Kennedy Gate
29. River Gate (seasonal)
30. Waring Gate (seasonal)

A Garden of Gardens

For the entertainment, recreation and instruction of the people.
—NYBG Charter

On a pleasant weekend, The New York Botanical Garden resembles nothing so much as a big family outing—a joyous extravaganza of people and plants, all lending their special color to the scene. Along the paths that wind through the Garden's 250 acres, couples wander, parents push strollers, children play. Amateur gardeners pause in front of plant beds, pondering new ideas to try out at home. Photographers snap pictures; artists paint. Older citizens rest on wayside benches, enjoying the day.

Many of NYBG's visitors come to admire the stunning seasonal displays, which range from bright waves of spring daffodils to the rich-hued tapestries of fall. For others, the Garden is simply a place of quiet beauty in which to retreat for a while from the city's grime and noise. For an increasing number, it is something else: a living museum of nature, an outdoor classroom in which almost anyone can learn a great deal about the world of plants.

Among the Garden's collections, trees and shrubs are dominant, as they are in nature, but here they are carefully selected and dramatized as exhibits in themselves. Thanks to the skills of NYBG horticulturists over the years, these major plantings also serve as backgrounds to enhance other floral displays, and as dense green buffers to screen the Garden from busy highways on all sides.

From the beginning, the basic framework was built around systematic collections of woody plants, some saved from existing specimens on the grounds, many gathered from various habitats in North America and abroad. They include locusts and other leguminous species; flowering magnolias, dogwoods, crab apples and cherries; rho-

During construction of the Rock Garden in the 1930s, workmen used a block and tackle to swing heavy stones into place. (NYBG Archive)

dodendrons and azaleas; oaks, maples and ashes; and a variety of coniferous evergreens, among them collections of pines, spruces and firs.

The most remarkable, and venerable, of the collections did not need planting; it has occupied its place of honor since long before the Garden was born. It is the 40-acre NYBG Forest, the only major patch of natural, uncut forest remaining from a vast woodland that once covered all of metropolitan New York. Located along a scenic, craggy gorge of the Bronx River near the center of the grounds, this unique relic—described in the early 1900s as the city's "most precious natural possession"—is one of the reasons the site was originally selected for a botanical garden.

For a long time it was called the "hemlock forest," but the dominance of eastern hemlock (*Tsuga canadensis*) has dwindled in recent decades; windstorms, insect infestations, polluted city air and soil compaction from generations of visitors' feet have all taken their toll. Though many fine hemlocks still stand, they share the spotlight today with deciduous trees, among them American beeches (*Fagus grandifolia*), red oaks (*Quercus rubra*) and tulip trees (*Liriodendron tulipifera*), plus other species that are sprouting up to change the character of the woods.

The forest and its environs still support a surprising variety of wildlife, including armies of squirrels and rabbits. Though they are generally more secretive or nocturnal, one can occasionally see raccoons, muskrats, skunks, flying squirrels, tiny voles and shrews. The forest has long been a favorite of bird watchers, who come to observe a great diversity of species, including such splendid sights as occasional black-crowned night herons, great blue herons and great horned owls. To try to save this fragile ecosystem for future generations, scientists of NYBG's Institute of Ecosystem Studies have launched a long-term study and management program (Chapter 7).

Along with its arboreal displays, NYBG boasts a score of gardens of various kinds, each organized around a collection of plants with a different educational theme. Among the most frequently visited, located on a rise bordering the forest, are the Rock and Native Plant Gardens, twin jewels in the Garden's crown.

The Thomas H. Everett Memorial Rock Garden, recognized as one of the finest public rock gardens in the United States, was the first project of its namesake, affectionately known to colleagues as "T. H." An experienced English horticulturist and a veteran of Kew, he was working on a private estate near New York when NYBG approached him to become its director of horticulture in 1932. Everett agreed, but made one condition of employment: that he be able to build a rock garden of his own. (A smaller Victorian rockery near Daffodil Hill, built in the Garden's earliest years, was deemed inferior and has long since been abandoned.)

Legend has it that T. H., a dapper, energetic gentleman—"He didn't walk, he ran," his secretary Lillian Weber recalls—designed his pet project on the back of an envelope while enjoying a proper gin and tonic at the end of the day. Choosing a depression between two natural rock outcroppings partially shaded by trees, he had a crew

A Tree-Lover's Walk

Some of the finest specimens at the Garden, and ones that are often overlooked, are the ornamental trees and shrubs right near the main entrance parking lots. Notable for their size, grace and horticultural merit, many were planted during the 36-year tenure of Thomas H. Everett as NYBG's director of horticulture from 1932 to 1968.

A good way to see them is to start with the *alleé* of tall tulip trees in front of the Museum Building, which have become a trademark of the Garden since their planting 90 years ago, then proceed in a circle as indicated on the diagram.

T. H. Everett Tizee and Shrub Walk Map

1. tulip tree, *Liriodendron tulipifera*
2. Glastonbury Thorn, *Crataegus monogyna* 'Biflora'
3. *Ilex opaca,* 'East Palatka'
4. Sargent's weeping hemlock, *Tsuga conadensis* 'Pendula'
5. Swiss mountain pine, *Pinus mugo* var. *mugo*
6. star magnolia, *Magnolia stellata*
7. pyramidal white pine, *Pinus strobus* 'Fastigiata'
8. bottlebrush buckeye, *Aesculus parviflora*
9. Japanese pagoda tree, *Styphnolobium japonicum (Sophora japonica)*
10. yew, *Taxus* cultivars
11. sugar maple, rock maple, hard maple, *Acer saccharum*
12. buttonwood, eastern sycamore, *Platanus occidentalis*
13. swamp white oak, *Quercus bicolor*
14. bur oak, mossy-cup oak, *Quercus macrocarpa*
15. British oak, English white oak, *Quercus robur*
16. Daimyo oak, *Quercus dentata*
17. green ash, *Fraxinus pennsylvanica*
18. Turkish hazelnut, *Corylus colurna*
19. Nikko fir, *Abies homolepis*
20. Norway spruce, *Picea abies*
21. London plane, *Platanus* × *acerifolia*
22. cork tree, *Phellodendron lavallei*
23. scarlet oak, *Quercus coccinea*
24. Kentucky coffee tree, *Gymnocladus dioicus*
25. pin oak, *Quercus palustris*
26. Chinese fringe-tree, *Chionanthus retusus*
27. white ash, *Fraxinus americana*
28. Oriental spruce, *Picea orientalis*
29. shadblow, Juneberry, *Amelanchier canadensis*
30. Colorado blue spruce, *Picea pungens* 'Glauca'
31. Wilson spruce, *Picea wilsonii*
32. Engelmann spruce, *Picea engelmannii*
33. Yeddo spruce, *Picea jezoensis*
34. paperbark maple, *Acer griseum*
35. silver Colorado spruce, *Picea pungens* 'Argentea'
36. tiger-tail spruce, *Picea torana*
37. Chinese stewartia, *Stewartia sinensis*
38. Black Hills spruce, *Picea glauca* 'Densata'
39. Austrian pine, *Pinus nigra*
40. Iigiri tree, *Idesia polycarpa*
41. silver maple, *Acer saccharinum*
42. blue ash, *Fraxinus quadrangulata*
43. Burkwood viburnum, *Viburnum* × *burkwoodii*

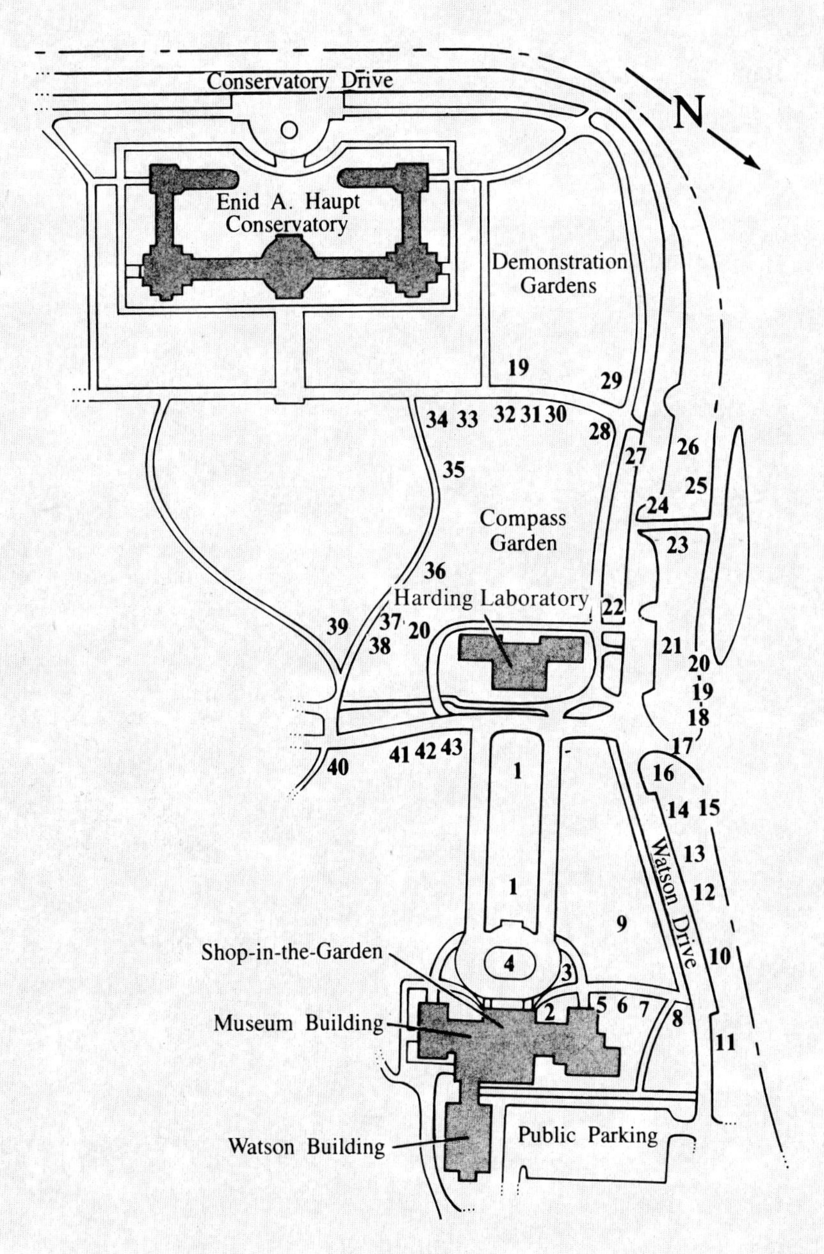

A cultivar of moss pink (Phlox subulata), *a popular Rock Garden plant.* (Allen Rokach)

of workers cart in additional rocks from elsewhere on the grounds, using a team of horses to haul them into place. Huge, flattish stones were set into the hillside around a new waterfall, which is artificial but so well designed that most visitors assume it is real.

In another ingenious piece of construction, shallow underground concrete basins were built into a rise across the way to provide water circulating under the roots of finicky alpine species, which were planted in gravel beds that simulate a glacial moraine. T. H. himself collected many of the garden's original plants during expeditions to the Rockies and southern Appalachians. Others, including finds from the Alps, the Pyrenees and the Himalayas, were donated by, or purchased from, other explorers, rock gardeners and nurseries.

Both the waterfall and the moraine were restored to proper working order in 1988. The Rock Garden today consists of close to a thousand species and varieties in several distinct collections, each illustrating the use of different plants in different habitats.

For visitors to the Rock Garden, the show starts as early as March with the first blooming of alpines in the moraine, among them silveredge primrose (*Primula marginata*), saxifrages like *Saxifraga* × *apiculata*, drabas (*Draba*), pinks (*Dianthus*), columbines (*Aquilegia*), gentians (*Gentiana*), rock jasmines (*Androsace*)—all with their roots kept cool by the circulating water so they can survive the unaccustomed lowland warmth.

On the opposite side of the path is a gravel scree, similar to the moraine but without water. Here Bob Bartolomei, the Rock Garden's curator, has planted alpines from drier zones, including the Mediterranean, Turkey and the American West. In wetter areas bordering a rivulet from the waterfall, moisture-loving primroses (*Primula*) put on a show in May.

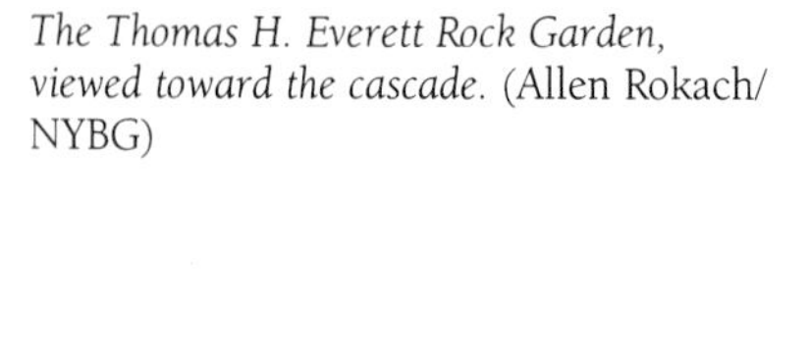

The Thomas H. Everett Rock Garden, viewed toward the cascade. (Allen Rokach/NYBG)

Japanese primroses (Primula japonica). (Allen Rokach/NYBG)

At the far end of the garden is a colorful area devoted to various heathers and heaths (*Ericaceae*), including a superb specimen of *Enkianthus perulatus*, a Japanese shrub that bears creamy white bell-like flowers in spring and turns a brilliant scarlet in fall. In the background is a China fir (*Cunninghamia lanceolata*), which provides a tall, striking foil with its bluish evergreen needles on branches that droop slightly at the tips.

Nearby is a small alpine meadow area planted with colorful spring bulbs—species of *Narcissus*, *Scilla* and *Crocus*, followed by autumn crocuses (*Colchicum autumnale*) and others that extend the display into fall.

"People love flowers, of course," says Bartolomei. "So we try to have something in bloom all season, not just in April and May. The challenge in rock gardening is that you are always trying to grow things where they really don't want to grow. If you can't imitate their natural environment—a high, cool mountaintop is pretty hard to duplicate in the Bronx—you try to compensate by providing water or shade."

The Rock Garden is a particular favorite among visitors to NYBG, he notes: "They think they are in the midst of something quite natural, which is the nicest compliment of all. A lot of them find it hard to believe that the rocks and plants here were carefully arranged by human hands."

Immediately adjacent to the Rock Garden is the equally distinguished Native Plant Garden, where species indigenous to the Northeastern United States are grown in

Pink and red Japanese primroses (Primula japonica) *along the Rock Garden's stream, with the white flower spikes of* Camassia esculenta *behind.* (Allen Rokach/NYBG)

Joe-pye weed (Eupatorium purpureum), *a native of marshy fields.* (Allen Rokach)

simulations of their own diverse habitats. The central section of the garden, bordering the small stream from the Rock Garden, is an open meadow rife with flowers and grasses that flourish in sunny, moist fields, putting on their most colorful show in late summer and fall. Among them are joe-pye weed (*Eupatorium purpureum*), boneset (*Eupatorium perfoliatum*), goldenrod (*Solidago*) and butterfly weed (*Asclepias tuberosa*). In wetter areas are New York ironweed (*Vernonia noveboracensis*) and the tall, spectacular Turk's-cap lily (*Lilium superbum*). A special section underlain with sandy, acidic soil illustrates plants common to the pine barrens of New Jersey and Long Island, including bearberry (*Arctostaphylos uva-ursi*), pitch pine (*Pinus rigida*) and sand myrtle (*Leiophyllum buxifolium*).

Wooded areas above the meadow are devoted to early blooming wildflowers: bloodroot (*Sanguinaria canadensis*), white trillium (*Trillium grandiforum*), pink and yellow lady's-slipper orchids (*Cypripedium acaule* and *C. calceolus*) and many more. An overhead misting system here can be turned on to water the plants and to cool, clean and moisten the air, making it possible to grow many delicate species that would not otherwise endure city conditions.

Notable in the woodland area is the F. Gordon Foster Hardy Fern Collection, a major display ground and experiment station for species that can be grown in northern regions. Established in 1986 with a gift of 70 kinds of ferns from Foster's own garden in New Jersey, it has been supplemented with others from the American Fern Society, and from John Mickel and Joseph Beitel, the Garden's fern specialists, who have collected many species from high-altitude areas of the tropical Americas that have also proved hardy here.

The collection today totals some 200 species and varieties of ferns, including both native types and ones from similar climates in the northwestern United States, northern Europe and Asia, the Himalayas, Japan, New Zealand and Latin America. Sections of acidic woodland are devoted to North American and foreign wood ferns (*Dryopteris*), lady ferns (*Athyrium*), holly ferns (*Polystichum*) and others. A small marshy area displays ferns that grow in acidic bogs, among them chain ferns (*Woodwardia* and *Lorinseria*) and royal ferns (*Osmunda regalis*). Lime-loving types like ostrich fern (*Matteucia struthiopteris*) and maidenhair ferns (*Adiantum*) are grouped in an area of limestone rocks.

Elsewhere, paths lead past more recent habitats developed by Juliet Alsop Hubbard, former curator of the Native Plant Garden, and her successor Kathryn Venezia. Particularly intriguing to home gardeners is a grouping of wild perennials that make fine garden subjects, including a small-petaled black-eyed susan (*Rudbeckia triloba*), New England aster (*Aster novae-angliae*), cardinal flower (*Lobelia cardinalis*), bee balm (*Monarda didyma*) and purple coneflower (*Echinacea purpurea*).

Prairie coneflower (Ratibida pinnata) *in the Native Plant Garden's meadow.* (Allen Rokach/NYBG)

Another area illustrates prairie plants once common on Long Island's Hempstead Plains: little bluestem grass (*Andropogon scoparius*) and bluecurls (*Trichostema dichotomum*). An alkaline habitat of serpentine rock brought in from Staten Island harbors species like moss pink (*Phlox subulata*) and harebell (*Campanula rotundifolia*). Still other areas illustrate plants native to various southeastern habitats that also thrive in more northerly climates, including various flowers, grasses and shrubs, as well as carnivorous types like sundews and pitcher plants (*Drosera* and *Sarracenia*). A small

Mayapple (Podophyllum peltatum) *in the woodland part of the Native Plant Garden.* (Allen Rokach)

bog features native fragrant water lilies (*Nymphaea odorata*), marsh marigolds (*Caltha palustris*) and aquatic irises like blue flag (*Iris versicolor*). A limestone area displays American twinleaf (*Jeffersonia diphylla*), shooting stars (*Dodecatheon*), crested irises (*Iris cristata*) and columbine (*Aquilegia canadensis*).

From the Rock and Native Plant Gardens, a path leads north to Rhododendron Valley, which displays many cultivars that were given to the Garden as seedlings in the 1940s by the estate of the renowned hybridizer Charles Owen Dexter, and that have now grown to substantial size. Particularly notable is the fragrant pink 'Scintillation', cuttings from which supplied the stock that has made this cultivar among the most widely sold rhododendrons today.

The valley also features the Murray Liasson Narcissus Collection. Some 15,000 bulbs of more than 80 types—arranged into 12 standard divisions according to flower size, proportion and color—not only provide striking spring displays but also serve gardeners as a useful demonstration of the many varieties hardy in the region.

Another path from the Rock and Native Plant Gardens, this one to the south, leads past a group of shrubby dogwoods (*Cornus*), whose yellow or reddish stems provide winter interest, then on to a swale where willows (*Salix*), bald cypress (*Taxodium distichum*) and other wetland species abound.

Here the trail branches left to the NYBG Forest, which can be entered here, or at a number of other points around its periphery. The forest, of course, deserves a special tour in itself.

To the right, a path named Azalea Way passes through the Sherman Baldwin Azalea Collection, which reaches its peak of color in May. Outstanding are the towering

Mixed Narcissus *varieties blanket a slope near Daffodil Hill.* (Allen Rokach/NYBG)

◀ *A brilliant red azalea,* Rhododendron stewartstonian, *on Azalea Way.*

The Peggy Rockefeller Rose Garden. (Allen Rokach/NYBG)

torch azaleas (*Rhododendron kaempferi*) in shades of pink, salmon and red; yellow azaleas (*R. lutescens*), Korean azaleas (*R. mucronatum*), pink-shell azaleas (*R. vaseyi*), orange flame azaleas (*R. calendulaceum* hybrids) and royal azaleas (*R. schlippenbachii*).

Other paths lead to the Donald Bruckmann Crab Apple Collection, where some 35 newer, improved varieties of *Malus* bloom in shades of white, pink and red from early spring to early summer. Some fine older crab apple varieties, as well as Asiatic azaleas and a collection of white pines, are displayed in other areas across the road. Adjoining the Bruckmann collection to the west is Daffodil Hill, where every April tens of thousands of daffodils of different varieties form bright blankets of color on the slopes.

Beyond a bridge over the Bronx River lies a pleasant walk through the Robert H. Montgomery Conifer Collection, started in 1947 when Colonel Montgomery, a New York lawyer, gave the Garden 200 specimens from his Connecticut estate that he had gathered during travels around the world. Here one finds many interesting species and varieties, including weeping hemlocks (*Tsuga canadensis pendula*), spreading types of English yew (*Taxus baccata*) and a fine specimen of weeping golden larch (*Pseudolarix amabilis nana*). Of special note is a grove of dawn redwoods (*Metasequoia glyptostroboides*), an unusual, deciduous conifer that was known only from fossils until 1945, when three living trees were discovered in China and propagated to restore the species to widespread use.

From the conifer collection, the main path leads to the two-acre Peggy Rockefeller Rose Garden, one of the newest and most visited of the gardens at NYBG. Originally designed for the site in 1916 by the noted landscape architect Beatrix Farrand, but abandoned for lack of funding after partial development, it was rebuilt in 1988 with a major gift from David Rockefeller in honor of his wife. The garden, with formal beds radiating out from a central wrought-iron gazebo, contains 2,700 plants of some 225 *Rosa* species and cultivars, arranged according to type, under the supervision of Michael Ruggiero, senior curator of the rose garden and other special gardens at NYBG.

A hundred varieties of old garden and shrub roses are planted in wide perimeter beds. These include bourbon, damask, gallica, centifolia, Portland, moss, hybrid musk and rugosa roses, among them the beautifully red-and-white-striped *Rosa gallica versicolor* (Rosa Mundi), the intensely fragrant damask rose 'Madame Hardy' and the raspberry pink rugosa 'Roseraie de l'Hay'.

The main central beds boast 130 kinds of modern roses, including such outstanding hybrid teas as the dark crimson 'Chrysler Imperial' and the pink 'Touch of Class'. Here, too, are many floribundas, crosses of hybrid tea roses and the smaller, profusely blooming polyanthas, among them 'Betty Prior' and 'Sunsprite'. Densely petaled grandifloras are represented by 'Queen Elizabeth', the first named grandiflora, and more recent cultivars like 'Tournament of Roses'. In arc-shaped beds around the gazebo, and along the perimeter iron fence, are minifloras and climbing roses, among them the yellowish and pink 'Spectra' and the light pink 'Clair Matin'.

Immediately to the north of the Rose Garden is the T. A. Havemeyer Lilac Collection, where 180 species and cultivars of lilacs (*Syringa*) bloom in late spring, their purple, lavender, pink and white flowers perfuming the air. The lilacs border an area where the Garden's former landowner, Pierre Lorillard, maintained his own "acre of roses" in the 1840s to supply rose petals for scenting the snuff he made in his nearby

A border of pink 'Betty Prior' floribundas in the Rockefeller Rose Garden. (Allen Rokach/NYBG)

Behind the scenes in NYBG's Propagation Range, a two-acre area of working greenhouses and nurseries, horticulturists tend tens of thousands of plants, like these caladiums, that are used in the Conservatory's changing exhibits and in outdoor displays. (The New York Botanical Garden)

Clematis *'Will Goodwin' on a trellis.* (The New York Botanical Garden)

mill. Now known as the Snuff Mill, this handsome fieldstone building has been designated a National Historic Landmark. It was restored in the 1950s as a public restaurant and outdoor café where visitors can retreat for lunch, and where meetings of various plant societies are often held. Opposite the lilac collection is Cherry Valley, where spring-blooming trees include rosebud cherries (*Prunus subhirtella*) and Japanese flowering cherries (*P. serrulata, P. yedoensis*).

At the crest of the hill is the Arlow B. Stout Daylily Garden, a collection of 120 daylily species and cultivars in a spectrum of colors. Commemorating Stout's outstanding achievements with the genus *Hemerocallis* while working as a botanist at NYBG, it includes many of his own cultivars, among them 'Theron', the first dark red daylily, which took Stout 25 years of crossing and recrossing to produce (see box). Also on display are all the past winners of the annual Stout Medal, the American Hemerocallis Society's highest award. Between the daylilies established in this garden and others planted around the grounds, all under the direction of NYBG gardener Gregory Piotrowski, NYBG displays some 1,300 cultivars by Stout and other hybridizers, many of them gifts from a single enthusiast, Rosewitha Waterman of Huntington, Long Island. Across the road from the daylily garden is the Children's Garden, a focal point of NYBG's educational program for youngsters (Chapter 6).

Gardeners of all ages who are looking for ideas find several gardens near the Conservatory well worth their time. Among the most instructive is the Jane Watson Irwin Perennial Garden, a gift of Helen Watson Buckner in memory of her sister. With areas recently renovated by New York garden designer Lynden Miller, it shows off a wide variety of handsome and durable varieties suitable for home use. Half the beds contain flowers whose tones are "cool"—blues, pinks, whites and grays—among them hostas, dicentras, phlox, aconitums and Japanese anemones. The other half has

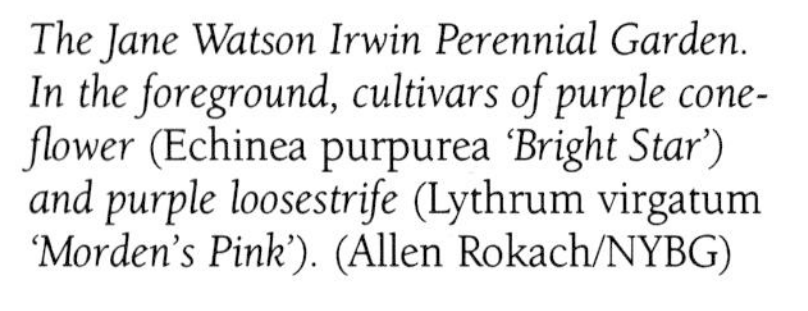

The Jane Watson Irwin Perennial Garden. In the foreground, cultivars of purple coneflower (Echinea purpurea *'Bright Star'*) *and purple loosestrife* (Lythrum virgatum *'Morden's Pink'*). (Allen Rokach/NYBG)

(Marcia Stevens/NYBG)

Chasing Rainbows

As a teenager growing up in Wisconsin in the early 1890s, Arlow Burdette Stout, known to his friends as "Bert," developed a consuming curiosity about the natural world. In observing plants and animals around his home, he noticed that an ordinary orange daylily his mother had planted near their porch failed to produce seed pods like other flowers. He wondered why.

Decades later, when Stout was achieving fame as NYBG's most prolific plant breeder and geneticist, he began to understand. Like many others of its kind, the tawny daylily, *Hemerocallis fulva* 'Europa', is limited by complex sterilities that reduce or prevent its ability to set seeds, though it is so indestructible, and multiplies so rapidly from spreading roots, that since its introduction to America by early settlers it has become a common sight on roadsides everywhere. Only through trial-and-error knowledge of the family's quirks had previous hybridizers been able to produce new strains at all.

Stout, if nothing else, was a patient man. Among his experiments, he made more than 7,000 cross-pollinations of 'Europa', from which he netted exactly 23 seed capsules containing 70 seeds. From these he raised 11 seedlings, then began to mate them with other daylilies, including the common lemon daylily, *H. liliaosphodelus* (*H. flava*), another favorite since colonial days.

A paved sitting area in the Irwin Garden, with crab apples (Malus) *in planters.* (Allen Rokach/NYBG)

After 25 years of painstaking work, which involved the crossing and recrossing of more than a dozen different species and selected cultivars, Stout produced a veritable rainbow of new colors. The climax was the first dark red daylily, a cultivar whose flowers shade from yellow throats into gorgeous mahogany petals that in some lights appear purplish black (shown in Eleanor Clarke's painting above*). He named his prize 'Theron' in honor of a member of the Garden's Advisory Council, Mrs. Theron G. Strong.

By the time he retired as NYBG's Curator of Education and Laboratories in 1947, Arlow Stout had produced a major pool of breeding stock for other hybridizers, who have created more than 30,000 daylily varieties, of which some 12,000 are on the market, and continue to register well over a thousand new varieties a year. Now available in an immense range of sizes, shapes and blooming times, and in every color except pure blue, the durable old daylily has become one of the most popular garden perennials since the days of the tulip craze three centuries ago.

varieties whose flowers are "hot"—red, orange and yellow, including crocosmias, euphorbias, heleniums, achilleas and kniphofias—blended with whites, grays and blues to unite the garden's two halves. In both areas are plants notable for their textures of green, gray or silver foliage. Forming an enclosure around one end are wooden arbors on which more than 60 species and cultivars of *Clematis* bloom freely from late spring until fall. To make the garden even more inviting to visitors, generous paved areas are broken up with small crab apples (*Malus*) in planters, and with handsome teak benches that provide places to sit.

Next to the Irwin garden is the Herb Garden, with brick paths laid in basketweave patterns and beds planted with some 60 species and varieties that illustrate traditional uses of herbs for fragrances, flavors, dyes and medicines. Nearby is a small Chemurgic Garden, devoted to other plants of economic significance. On higher ground bordering the Conservatory is a substantial collection of irises planted in honor of Carl Totemeier, former vice president for horticulture at NYBG.

At the opposite end of the Conservatory, the Barbara Foster Vietor Memorial Walk is a favorite among visitors for its ornamental bulbs and bedding plants. Spring features 120 different kinds of tulips arranged in 13 divisions from single earlies to later doubles, fringed, parrot and lily-flowered types. Displays continue through fall with colorful annuals planted as a "reference library" for gardeners, including cleomes, begonias, impatiens, zinnias and marigolds. Senior curator Mike Ruggiero, who plans the displays, includes award-winners and other worthy cultivars that come on the market each year.

The Irwin Garden's shady section: Hosta *'Blue Cadet' blooming in front,* Astilbe *'Peach Blossom' in back.* (Michael Ruggiero)

Adjoining the Vietor Walk, in an area surrounded by part of NYBG's extensive fir collection, are five Demonstration Gardens, installed in recent years to provide ideas for home plots.

Helen's Garden, a gift in memory of Helen Goodhart Altschul, displays plants notable for their fragrance, selected so that there is always something noteworthy in bloom. The Mae L. Wien Summer Flower Garden features plants that furnish cut flowers through the warmer months. The Louise Loeb Vegetable Garden shows home gardeners how to plan a succession of food crops that are as decorative as they are edible, using raised planting beds that provide good drainage and easy maintenance, as well as wooden trellises and fences on which pole beans, peas, cucumbers and small melons are trained to save space and intensify yields.

Cosmos *'Radiance' in the Mae L. Wien Summer Flower Garden.* (Allen Rokach)

A Calendar of Color

	JANUARY	FEBRUARY	MARCH	APRIL	MAY	JUNE	JULY	AUGUST	SEPTEMBER	OCTOBER	NOVEMBER	DECEMBER
Major seasonal displays												
Enid A. Haupt Conservatory	■	■	■	■	■	■	■	■	■	■	■	■
Orchid Rotunda (Museum Building)	■	■	■	■	■	■	■	■	■	■	■	■
Everett Rock Garden			■	■	■	■	■	■	■	■	■	
Native Plant Garden				■	■	■	■	■	■	■	■	
Rockefeller Rose Garden					■	■	■	■	■	■	■	
Demonstration Gardens												
White Country Garden			■	■	■	■	■	■	■	■	■	
Loeb Vegetable Garden				■	■	■	■	■	■	■	■	
Helen's Garden				■	■	■	■	■	■	■	■	
Wien Summer Garden					■	■	■	■	■	■	■	
Bryce Wildlife Garden				■	■	■	■	■	■	■	■	■
Stout Daylily Garden					■	■	■	■	■			
Daylily-Daffodil Walk				■		■	■	■	■			
Vietor Memorial Walk						■	■	■	■	■		
Notable species displays												
Wintersweet Conservatory Courtyard	■	■	■									
Witch hazels Conservatory Drive	■	■	■									
Winter jasmine Museum Building	■	■	■	■								
Pussy willows Swale		■	■									
Cornelian cherries Conservatory Drive			■	■								
Andromedas Conservatory Drive		■	■	■	■							
Winter hazels Garden Way		■	■	■								

WHILE SPRING'S SPECTACULAR BURSTS OF FLOWERS DRAW APPRECIATIVE CROWDS, OTHER FLORAL displays make a visit to the Garden worthwhile at any time of year. Locations of notable gardens and species are indicated on the accompanying map of the grounds. (Due to weather, blooming periods may vary by as much as two weeks. Consult NYBG for dates of special flower festivals, Conservatory shows and other events.)

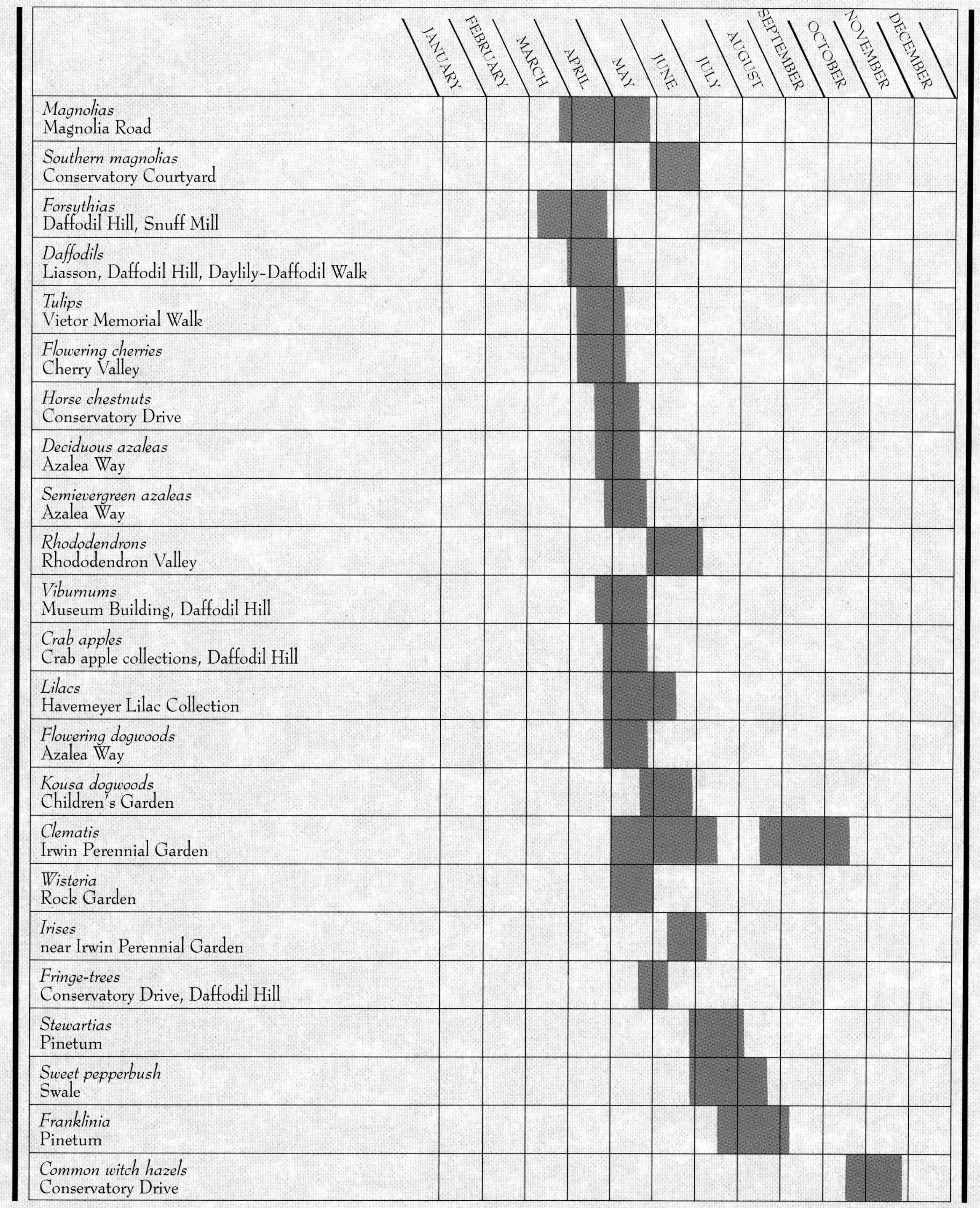

	JANUARY	FEBRUARY	MARCH	APRIL	MAY	JUNE	JULY	AUGUST	SEPTEMBER	OCTOBER	NOVEMBER	DECEMBER
Magnolias Magnolia Road			■	■	■							
Southern magnolias Conservatory Courtyard						■						
Forsythias Daffodil Hill, Snuff Mill			■	■								
Daffodils Liasson, Daffodil Hill, Daylily-Daffodil Walk				■								
Tulips Vietor Memorial Walk				■	■							
Flowering cherries Cherry Valley				■	■							
Horse chestnuts Conservatory Drive				■	■							
Deciduous azaleas Azalea Way				■	■							
Semievergreen azaleas Azalea Way				■	■							
Rhododendrons Rhododendron Valley						■						
Viburnums Museum Building, Daffodil Hill				■	■							
Crab apples Crab apple collections, Daffodil Hill				■	■							
Lilacs Havemeyer Lilac Collection				■	■	■						
Flowering dogwoods Azalea Way				■	■							
Kousa dogwoods Children's Garden					■	■						
Clematis Irwin Perennial Garden					■	■	■		■	■	■	
Wisteria Rock Garden					■							
Irises near Irwin Perennial Garden						■	■					
Fringe-trees Conservatory Drive, Daffodil Hill					■	■						
Stewartias Pinetum							■	■				
Sweet pepperbush Swale							■	■				
Franklinia Pinetum							■	■	■			
Common witch hazels Conservatory Drive										■	■	

The showy flower heads of wild leek (Allium ampeloprasum) *and borders of sweet alyssum* (Lobularia maritima) *in the Loeb Vegetable Garden.* (Allen Rokach/NYBG)

The Rodney White Country Garden celebrates the northeastern countryside with more than a hundred kinds of flowers, shrubs and trees that can bring the native beauty of woodlands and meadows into backyards. Newest of the demonstration gardens is the Della J. Bryce Wildlife Garden, designed to attract birds, butterflies, bees and other creatures, and to offer visual interest year-round.

Across the way from these gardens are beds of peonies with showy, fragrant blooms, including early, midseason and late types in shades of white, pink and red. Beyond them a curving walk is bordered by plantings of 160 daffodil cultivars, interplanted with 160 kinds of daylilies to provide color from early spring through fall. The walk also leads past part of NYBG's spruce collection to the Armand G. Erpf Compass Garden, which depicts the points of the compass with hedges and seasonal flowers in Victorian style.

Each of NYBG's gardens has a professional gardener or curator assigned to it to maintain current plantings and introduce new ones. Several of the larger gardens also have volunteer committees to help guide policies and acquisitions and raise funds for their own small endowments. Each of these committees, in turn, reports to an overall horticultural committee of NYBG's board.

Those who care for the Garden are mindful of the nature of their custodianship. "We have a unique piece of property here, right in the middle of the city," says Carl Totemeier, who retired in 1990 after a distinguished record in reinvigorating NYBG's collections with the help of his horticultural director Richard Schnall. "It has all the things some people would pay a fortune to create—interesting, undulating terrain, a river, a natural forest, rock outcroppings, wetlands, open fields, a pair of lakes. Whatever is done in the future should be designed to enhance that setting, not detract from it."

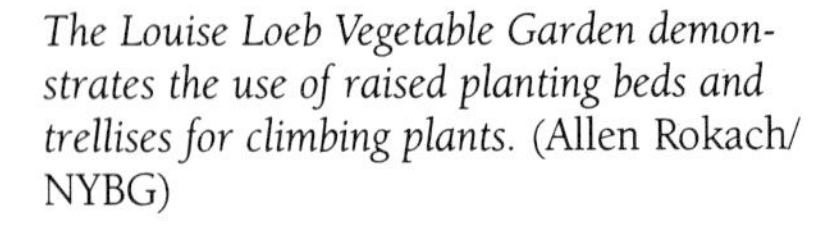

The Louise Loeb Vegetable Garden demonstrates the use of raised planting beds and trellises for climbing plants. (Allen Rokach/NYBG)

To further that goal, the entire Garden has been surveyed and laid out on a grid of 50-meter squares, allowing detailed, ongoing inventories of the terrain and major plantings with the help of aerial photographs. Under the eye of Bruce Riggs, manager of horticultural interpretation and plant records, data can be fed into a computer system to evaluate the various collections and, when necessary, bring them up to date. Trees, for example, will be rated by age and condition as a guide to maintenance and replacement. Species that have not proved worthy of interest, or adaptable to the region, will be removed and other, more appropriate ones introduced.

Other moves have been aimed at making the Garden more inviting and informative. Visitors arriving at the main gate now leave their cars in the parking lots; except for maintenance vehicles, automobiles have been banned from the inner roads to make the grounds safer and more pleasant for walking. In the Museum Building, near an expanded Shop in the Garden that sells plants, books and other horticultural wares, visitors can stop at an orientation desk to plan their outings, or pause for refreshments at a terrace café outside before boarding trams that take them on tours of the grounds.

DCCCXCIX
ENID A. HAUPT CONSERVATORY

New York's Crystal Palace

"Presto! It is a thing of beauty and a joy forever."

For more than 90 years, the centerpiece and symbol of The New York Botanical Garden has been its Conservatory, an elegant wedding cake of a building that rises above the surrounding greenery like a vision of yesteryear. As a structure, it is not only a grand piece of nostalgia and an architectural delight. It also houses some of the country's finest collections of tropical, subtropical and desert flora, as well as seasonal exhibits and flower shows that are among the best-attended horticultural events in the Northeast.

The building, like other "glass houses" of its era, was inspired by the nineteenth-century passion for collecting exotic plants from the tropics and raising them indoors. Among the early Victorian prototypes that influenced its design, and that of many others before it, was the famous Crystal Palace designed by Joseph Paxton and built for London's Great Exhibition of 1851 and since destroyed. A more immediate model—particularly noticeable in the graceful curves of the Conservatory's end pavilions—was a favorite of Nathaniel Lord Britton, the first director, and his wife: Decimus Burton's classic Palm House of the Royal Botanic Gardens at Kew, still one of the most beautiful greenhouses in the world.

The Conservatory was not the first major glass structure erected in America for the public display of plants. That distinction belongs to the spectacularly ornate Horticultural Hall built for the Philadelphia Centennial Exposition of 1876, which set off a flurry of new conservatories across the country sponsored by proud city fathers and philanthropists (many, like Philadelphia's, have been torn down, largely due to high costs of maintenance). Nevertheless, NYBG's entry has been described by British gar-

◀ *Illuminated for special occasions, the Enid A. Haupt Conservatory takes on a magical quality after dark.* (Allen Rokach/NYBG)

den writer May Woods as "the premier botanical conservatory in America," an assessment with which Garden enthusiasts tend to agree.

Designed by New York architect William R. Cobb, the structure was fabricated by the leading greenhouse firm of Lord & Burnham, with the help of Hitchings & Company, and built by John R. Sheehan under contract with New York City's Department of Parks. The building was laid out in a rectilinear C shape more than 500 feet wide, consisting of 11 interconnected pavilions embracing an entrance court with reflecting pools on either side. (It is said that Britton wanted to extend the pavilions to complete a square, with a second dome opposite the first, but city officials apparently felt enough was enough.)

The airy iron framework, started in 1899, was covered with 17,000 individual panes of glass, providing almost an acre of enclosed growing space. By the time of the grand opening of most of the pavilions in June 1900, they had been crammed with 9,000 plants, many of which had been accumulating in the greenhouses of Columbia College as a temporary home. Though the board had been ready to allot $10,000 for the displays, so many gifts came in from patrons, large and small, that the frugal Britton boasted he had to spend only $100 of NYBG funds.

The New York Times reported that thousands of people came to see the brand new Conservatory on a summer Sunday when the weather was hot, and considerably hotter indoors with the sun shining through the glass:

"The gardeners opened the doors and Presto!, the inside is a thing of beauty and a joy forever, filled with beautiful and rare plants. . . . There have been many gifts, for people are being most generous to the Garden, florists as well as private individuals."

A large pool in the central courtyard displays tropical water lilies, (Nymphaea), *lotus* (Nelumbo) *and other aquatic plants. At center are the tray-like leaves of the giant water lily* Victoria amazonica. (The New York Botanical Garden)

▶
Looking up into the central dome at night.

In the central palm dome, it was noted, there were "five large, tall palms, given this last week by Miss Helen Gould . . . from 15 to 20 years old and so large they have outgrown Miss Gould's palm house, where they were threatening to lift the roof." One of the Garden's most generous benefactors was Mrs. Oakes Ames of Massachusetts, who contributed a "wonderfully great anthurium" as well as tree ferns, orchids and "an excellent collection of pitcher plants." In 1908 the entire Ames collection of orchids—1,500 from all over the world—was bequeathed to the Garden and created an even greater sensation when put on display.

Mrs. George Such of South Amboy gave a large number of lady's-slipper orchids from the East Indies, and John Crosby Brown parted with a large, spreading pawpaw tree. In an especially neighborly gesture, a Mrs. Mace, who lived near the Garden, donated "a remarkably large rubber plant, at least 10 feet high." Most of the presents thrived in their new home, so well, in fact, that some years later a vigorous palm began to lift out panes in the 90-foot-high central dome and had to be removed (perhaps one of Miss Gould's, up to its old habits).

New York's new crystal palace soon became an attraction for visitors from near and far. Among them were swarms of schoolchildren, who trooped in dutifully to gaze at strange, leafy specimens they had only read about in books. Even more popular among their elders were the lavish displays of flowers, including formal banks of tulips and daffodils brought into early bloom indoors, a tradition that continues in the Conservatory's seasonal shows today.

A tropical water lily in bloom. (Allen Rokach/NYBG)

Like other owners of large glass structures, however, NYBG gradually became aware of certain facts of life. In 1901, it was reported with relief that the Conservatory weathered a "moderately severe hailstorm, the hail stones glancing from the curved glass surfaces without fracturing a single pane." But over the years, rusting and rotting from moisture, inside as well as out—combined with the normal glass breakage and leaky roofs to which all greenhouses are prone—led to a need for constant maintenance and repair. As time went by, antiquated heating and ventilating systems worked erratically and occasionally not at all, threatening the survival of the plants.

Azaleas and primroses in a spring exhibit. (Allen Rokach/NYBG)

Periodic renovations were made, including major ones in 1935 and 1950, during which, unfortunately, much of the building's fine old ornamentation was removed. By 1972, following a cutback in city funds, the Conservatory was in such sorry shape it became evident that it would have to be totally rebuilt or torn down. A $2.5 million fund-raising campaign was launched, with the city agreeing to meet up to $1.5 million of the project's costs, then estimated at $3.5 million plus. In 1974 the Conservatory was designated an official city landmark, but the city itself, on the edge of bankruptcy, announced that it had to defer capital expenditures of any kind. By then further deterioration, combined with an inflationary economy, was driving costs through the building's fragile roof.

With more courage than confidence, the Garden's board of managers decided to go it alone. On May 1, 1975, a fund-raising ball for 450 selected guests was held in the Conservatory—during which, of course, it rained, and workmen were summoned hastily to cover offending leaks with plastic sheets ("When Doris Duke gets rained on, you get a little worried," observed Carlton Lees, the Garden's senior vice president in charge of the restoration). Nevertheless, one of the guests, Mrs. Paul Mellon, arranged for a gift of $500,000 from the Mellon Foundation, and others pledged smaller amounts. But that left a long way to go.

At this point, Enid Annenberg Haupt stepped in. A garden lover, philanthropist and board member of NYBG—as well as heiress to a publishing fortune and former publisher and editor of *Seventeen* magazine—Mrs. Haupt offered to redo the Palm Court and its dome to the tune of $850,000. Then one day, when the goal still seemed unattainable, she invited Lees and other Garden officials to her Park Avenue apartment for lunch.

Lees recalled that "everyone was walking on eggs," but that in the course of the conversation their hostess casually said that she wanted to cover the whole thing. ("When we got outside," said Lees, "we turned and looked at each other; we still couldn't believe our ears"). Mrs. Haupt was to increase her contribution to $5 million, and later add another $5 million as an endowment for maintenance.

Since the original blueprints had been lost, early photographs of the building had to be used as a guide. With the help of nineteenth-century restoration specialists, the needed drawings were made, translated into ornate floral patterns by a woodcarver

Chrysanthemums in myriad varieties and arrangements highlight a fall display.
(Allen Rokach/NYBG)

and then cast in rust-proof aluminum to replace the iron decoration that had been removed.

On March 13, 1978, the aging crystal palace, returned to its original splendor,* was reopened by its benefactor, who snipped a lavender ribbon to the strains of a string ensemble and stepped inside. At the ceremony that followed, the building was named for her.

Visitors to the Enid A. Haupt Conservatory today generally enter on the east side of the central dome, which is the most easily accessible to the rest of the Garden. On the west side is the original, formal entrance, where a handsome *alleé* of pleached European hornbeams (*Carpinus betulus*) borders a courtyard that has central beds of Japanese boxwood (*Buxus microphylla japonica* 'Green Beauty'). Flanking the courtyard, a Temperate Pool displays hardy water lilies (*Nymphaea*), cattails (*Typha*), water-tolerant irises (*Iris pseudocaria*) and other aquatic plants; a Tropical Pool is filled with various species of tropical water lilies, including the giant *Victoria amazonica*, as well as the sacred lotus (*Nelumbo*).

The Palm Court under the Conservatory's dome, a circular structure 100 feet across and 90 feet high, houses NYBG's extensive collection of tropical palms. To visitors who think of these trees only as exotic ornaments, the exhibit points out that native cultures around the world depend on them for a host of products, including oil, food, drink, fibers, construction materials and fuel (see Chapter 7).

Among the most useful is the coconut palm (*Cocos nucifera*), a species so widely cultivated it is estimated that there is the equivalent of a tree for every family on earth. A particularly arresting specimen is the sugar palm (*Arenga pinnata*) from Malaya, whose flower clusters produce a sap that is boiled into sugar, and which bears some

Visitors admire displays at the Conservatory's annual spring show.
(Allen Rokach/NYBG)

*At this writing, plans are being drawn up for another needed renovation of the Conservatory, pending allocation of the necessary city funds.

A Japanese Kiku chrysanthemum of the "spider" type. (Allen Rokach/NYBG)

of the largest leaves in the plant kingdom, each 20 to 30 feet long. There are also tall date palms (*Phoenix dactylifera*) from West Asia and North Africa, a species believed to have been in cultivation for 8,000 years; desert fan palms (*Washingtonia filifera*), the only palm genus native to California (also known as the petticoat palm for the skirt of dead leaves covering its upper trunk); and the Mexican fan palm (*Washingtonia robusta*), which in its native setting commonly exceeds a height of 100 feet.

On either side of the central dome, the Conservatory's galleries are devoted to different functions. The five greenhouses on one side contain the Garden's permanent collections of tropical, subtropical and desert plants. The five on the other side accommodate largely changing displays, which are set against semipermanent backdrops of trees and shrubs from subtropical and warm temperate zones.

Notable among these exhibits are NYBG's seasonal shows, which draw tens of thousands of visitors each year. From early February to mid-April, winter-weary guests are treated to a foretaste of spring: daffodils, tulips, hyacinths and other bulbs brought into bloom well ahead of their outdoor flowering time in NYBG's Propagation Range. In late spring the bulbs are replaced by colorful perennials and annuals. In summer the displays run to begonias—of which NYBG boasts one of the finest collections in the U.S.—as well as caladiums and other tropical species that can stand the summer warmth, often designed around a theme such as Spanish or Polynesian gardens.

Autumn brings arrays of fall-blooming chrysanthemums, both the familiar hardy species and the spectacular Japanese *Kiku* types. The latter, among the oldest ornamental plants in cultivation, are arranged in traditional bush and cone styles, as well as pagodas, trees, columns and small, medium and giant cascades. From late Novem-

Red and white poinsettias lead to a decorated tree at the Christmas show. (Allen Rokach/NYBG)

ber to early January, when the Garden puts on its Holiday Show, the galleries glow with red, white and pink poinsettias, amaryllises, cyclamens, Eucharist lilies, kalanchoes and paperwhite narcissus, which form a backdrop for gaily decorated wreaths and trees. Adding to the festivity are caroling choirs, ballet performances, demonstrations of wreath-making and holiday cooking, storytelling, magic shows and a Santa Claus.

In addition to seasonal displays, the Conservatory's galleries are the scene of special horticultural exhibitions. A longtime favorite is the Greater New York Orchid Show, staged in spring in association with the Greater New York Orchid Society. As many as four pavilions are turned over to 70 or more exhibitors from around the world, who display several thousand varieties, including the latest prize-winning cultivars. Plants from the Garden's own orchid collection are also displayed at the show, and at other times throughout the Conservatory as they come into bloom. Another popular event is the Bonsai Show, held in cooperation with the Yama-ki Bonsai Society in early fall, when more than a hundred specimens of tiny trees from leading private collections, as well as the Garden's own, are put on view.

After normal visitors' hours, the Conservatory periodically serves as a setting for a variety of glittering private gatherings. On an evening in late May or early June, it is the scene of the Garden's Spring Party, an annual fund-raising ball organized by volunteer committees. In recent years the ball has become a leading event on the New York social calendar, with as many as 750 patrons dining and dancing under a large tent set up temporarily in back of the dome. Later in June comes the Founders' Award Dinner, another NYBG fund-raiser at which corporations sponsor tables to honor an outstanding business leader.

At various times during the year the Conservatory is also rented out as a dramatic backdrop for corporate receptions, weddings and other parties. These are held in a

▶
In the New World Desert gallery, a dazzling variety of cacti. (Allen Rokach/NYBG)

REL CACTUS
TUS GRUSONII
ACEAE
RAL MEXICO

gallery that also serves to show off tropical species in hanging baskets and containers, as well as in an adjoining gallery called the Orangerie, where citrus and other species are planted in tubs so that they can be easily moved around. Guests are free to wander through all the Conservatory's galleries, which take on a special magic when the entire structure and its plantings are lighted after dark.

Beyond the Orangerie, the last gallery on this side of the building is devoted to economically useful plants, including many used for food or flavoring: pineapple, mango, banana, papaya, avocado, passion fruit, guava, ginger, cinnamon, allspice, sugar cane, rice and yams. Medicinal species include cinchona, the source of quinine, and Madagascar periwinkle, from which chemicals to treat leukemia have been derived. From this gallery, visitors enter a tunnel that was built under the entrance courtyard during the 1970s renovation, providing a second link between the two arms of the Conservatory, as well as space for graphic exhibits along its walls.

(Marcia Stevens/NYBG)

Armed and Dangerous

To many visitors, the Conservatory's most extraordinary displays are its collections of cacti and other spiky, succulent species from arid regions of the world. "Armed and dangerous" reads a sign at the gallery entrance, warning the unwary against painful contacts, while adding to the aura of mystery that surrounds these bizarre but often strangely beautiful plants.

Naomi Barotz, the collection's curator, explains that the prickly weapons and barrel-like shapes weren't developed out of sheer nastiness, but are ingenious adaptations to a harsh desert environment in which few plants can survive. Instead of conventional leaves, which would transpire precious water, cacti rely for photosynthesis on their fleshy green stems, whose waxy coatings retard evaporation from the water-storing tissues inside. Sharp spines—and in some species, a whole fuzzy overcoat of painfully barbed hairs—serve a dual purpose: they help reduce water consumption by shading the plants and decreasing evaporation from wind, and at the same time deter hungry, thirsty desert animals from eating them. To collect all the water they can during infrequent rains, cacti and other succulents have unusually wide, shallow roots, which may spread 10 feet on a plant only 3 feet high. Even more astonishing on such fierce-looking plants are the delicately tinted flowers they produce, luring pollinators into perpetuating their race.

The Garden's cactus collection had its origins in the early 1900s, when Director Nathaniel Lord Britton organized a series of expeditions to document a plant family about which relatively little was known. With another botanist, Joseph Rose, he studied and collected in the West Indies and the American Southwest, dispatching Rose for further explorations in South America from 1912 to 1917.

The result of their labors was *The Cactaceae*, a four-volume treatise illustrated by the gifted botanical artist Mary Eaton. It was published in 1920 by the Carnegie Institution, whose founder, Andrew Carnegie, had helped start the Garden and supported its scientific efforts until his death in 1919.

One of the plants Britton and Rose classified on their travels was the giant saguaro of Arizona and Mexico, which rises to a height of 30 or 40 feet and branches like a candelabra with age, bearing flowers at the tops of its stems in late spring (shown close up in Mary Eaton's painting). In honor of their patron, they named the species *Carnegiea gigantea* (other botanists classify it as *Cereus giganteus*), and arranged for a specimen to be brought back in their collections for display.

Though the original has since been replaced by another plant, the big cactus still dominates the Conservatory's New World Desert exhibit like an exclamation point—a salute to a giant among the Garden's friends.

On the far side of the tunnel are the Conservatory's permanent plant collections. First is the New World Desert gallery, which displays cacti and succulents from arid regions of North, Central and South America. The fascinating, often weird, appearance of many of these species, which makes them favorites for children's tours, are a result of adaptations to temperatures reaching 120 degrees and rainfall of only a few inches a year.

Among the most impressive specimens in the collection, Curator Naomi Barotz points out, is a giant saguaro cactus, a species named *Carnegiea gigantea* in honor of Andrew Carnegie, one of NYBG's founders (see box). Near the specimen, which is more than a century old, is the brown skeleton of another saguaro that reveals the unusual woody structure of these plants. Another prize is a large specimen of the rare boojum tree *(Fouquieria columnaris)*, found only in a small area of Baja California. Also on display are many other kinds of cacti, as well as agaves, aloes, yuccas, paloverdes, acacias and ocotillos. Adding dashes of color are desert wildflowers like California poppies (*Eschscholzia californica*) that bear brilliant orange flowers in spring.

Different ways of adapting to similar dry conditions, which botanists term "convergent evolution," are evident in an adjoining gallery devoted to species of the Old World Desert, the arid regions of Europe, Africa, Asia and Australia. In addition to cacti and succulent species are many varieties of South African bulbs, which flower in fall and winter in the Conservatory (as opposed to spring and summer in their native Southern Hemisphere).

Arching fern fronds and overlapping leaflets create delicate patterns. (The New York Botanical Garden)

The next gallery exhibits plants that grow mainly in subtropical regions, including Southern California, Africa, Australia and the Mediterranean. Of particular interest are two special collections of smaller plants housed in transparent cases of their own. One is of exotic orchids, displayed on a rotating basis as they come into bloom. The other is of carnivorous species, which rarely fail to get a reaction when visitors find out how they work. Displayed in a simulated bog are Venus flytraps (*Dionaea muscipula*), pitcher plants (*Sarracenia*), sundews (*Drosera*) and bladderworts (*Utricularia*), all of which have ingenious methods of luring and trapping insects to supplement a basic lack of nutrients in their acidic habitat.

Even more memorable for many visitors is the adjoining gallery, which brings an exotic jungle to mind. Called the Fern Forest, it is home to NYBG's outstanding collection of tropical ferns, which have been gathered from warm regions around the world. Many of their natural habitats are simulated in the forest, which has a "skywalk" that enables visitors to observe the plants from various angles and close up.

Here hundreds of species are arranged by families, in rough evolutionary order, explains Mobee Weinstein, the collection's caretaker. In corners flanking the entrance are rare ancient ferns like *Angiopteris palmaformis* from Southeast Asia, with leaves over 10 feet long; *Marattia* from Mexico and New Guinea; fern allies like clubmosses *(Lycopodium)* from the Old and New World tropics. Nearby are beds of maidenhair ferns (*Adiantum*), spleenworts (*Asplenium*) and brake ferns (*Pteris*). Hart's-tongue ferns (*Phyllitis*) and other lime-loving types are grown in a small limestone rockery. Staghorn ferns (*Platycerium*), named for their resemblance to antlers, cling to trees as they do in their native habitats.

Besides ferns, other tropical species on display include such useful ones as rubber plants (*Ficus elastica*) and mahogany trees (*Swietenia mahogani*). The oldest and largest

◄ *In the Fern Forest, tropical species are grouped around a cascade.* (Allen Rokach/ NYBG)

Oval-leaved Salvinia, *among the smallest members of the fern family, floats in a water-filled urn.* (Allen Rokach)

specimens here are a huge kapok tree (*Ceiba pentandra*), a species whose silky fibers are a traditional stuffing for life preservers, mattresses and sleeping bags, and an African nutmeg tree (*Monodora myristica*), whose fruits are used for flavoring. Both were donated around the time of the Conservatory's completion in 1901.

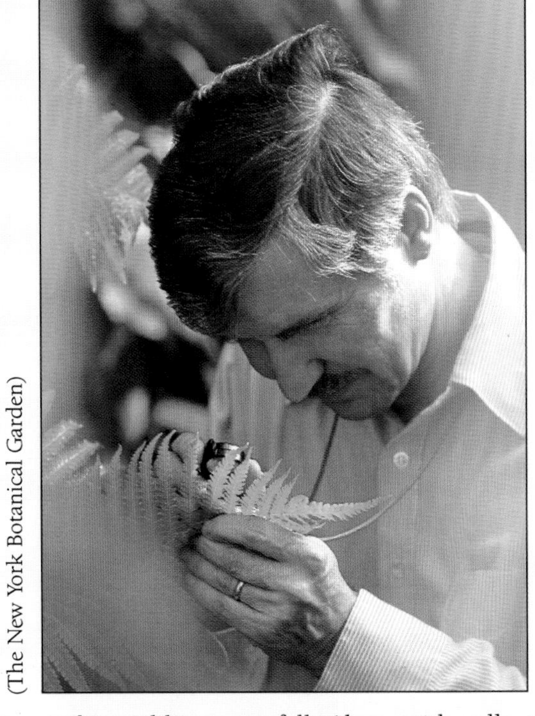

(The New York Botanical Garden)

A Fascination with Ferns

A POPULAR YEAR-round exhibit in the Conservatory is the Fern Forest, which contains the largest selection of ferns under glass in the United States—some 500 kinds displayed around a trickling waterfall. Along with collections of hardy ferns grown in the Native Plant Garden, it also represents a major resource for education and research.

Presiding over this fronded empire is John Mickel, NYBG's senior curator of ferns (above). Mickel is recognized around the world as an authority on the pteridophytes, embracing 12,000 species of ferns, as well as equally primitive clubmosses, spikemosses, quillworts and horsetails, which, like ferns, have no flowers and reproduce by spores instead of seeds. With co-author Joseph Beitel, another fern specialist and the Garden's curator of plant records, Mickel recently completed a comprehensive study of the fern flora of Oaxaca, an area of Mexico to which he conducts regular NYBG tours for fern enthusiasts. For broader audiences, he has written *The Home Gardener's Book of Ferns*.

One of the fascinations of ferns, Mickel will tell you, is that, except for algae and bacteria, they are the oldest living plants. Emerging from the prehistoric ooze 400 million years ago, pteridophytes were the first to develop vascular systems to draw up nourishment, creating dense blankets of vegetation that would eventually decay and be compressed into the world's deposits of coal. In the process of adapting to diverse environments—species grow everywhere except the coldest and driest regions—the plants have themselves become almost incredibly diverse, ranging from aquatic mosquito ferns less than an inch across to giant tree ferns 60 feet tall.

In his travels, John Mickel has discovered and classified many species that were previously unknown. What gives him even greater pleasure, however, is finding "some wonderful new plant from a Mexican mountaintop" that can add new beauty to people's homes.

Two specimens of Carnegiea gigantea *in the New World Desert exhibit, including a dead one showing the species' woody structure.* (The New York Botanical Garden)

During the 1970s renovation, the gallery was given a central "mountain" of lightweight lava rock incorporating a waterfall and pool, over which the aerial walkway was built. Rising around and above it are towering tree ferns (*Dicksonia, Cibotium* and *Cyathea*) from Australia, Hawaii, Mexico and the West Indies, whose huge fiddleheads and fountaining fronds can be seen at close range from the ramp. Providing a striking contrast in size are water-filled urns containing water spangles (*Salvinia*), prolific, free-floating plants whose leaves are less than an inch across. Even tinier mosquito ferns (*Azolla filiculoides*) float on the pool below the cascade.

Beyond the Fern Forest, the last stop is a gallery devoted to tropical flora. Of major interest here is NYBG's outstanding collection of cycads, large, primitive, palmlike plants that date back to the times of the dinosaurs and are now listed as endangered species wherever they still exist in the wild. Bromeliads, Chinese evergreens, dieffenbachias and other tropical species commonly grown as houseplants are also displayed. Favorites among visitors are a "chocolate tree" (*Theobroma cacao*) from Central and South America whose seeds are the source of flavoring for cocoa and chocolate; dwarf bananas (*Musa acuminata*); and the "Panama hat plant" (*Carludovica palmata*), from whose leaf fibers the hats of that name are made.

Climbing the sides of the gallery are golden trumpet vines (*Allamanda cathartica*) with bold yellow blooms. Flowering in the dead of winter when the landscape outside is largely bare, they and the other colorful tropicals are what many visitors remember most vividly after a tour through the crystal palace in the Bronx.

(The New York Botanical Garden)

The Orchid Man

Keith Lloyd, curator of NYBG's Sarah Davis Smith Orchid Collection, tends 5,000 plants of 650 species in five greenhouses of the Propagation Range. As they come into bloom, he selects outstanding ones for display in the Conservatory, and in the Museum Building's rotunda. Many of his showy cattleyas and delicate phalaenopses come from Brazil, which, appropriately, is a focus of NYBG's scientific research.

Born in Jamaica, Keith immigrated to the United States as a teenager in 1957 and subsequently served as an Air Force medic, caring for wounded servicemen during the Vietnam War. On his return to New York City, he became an enthusiastic volunteer in a community garden on a vacant East Side lot, growing vegetables and flowers before going to work as a nurse at the Lenox Hill Hospital on the evening shift. His love of plants finally prompted him to enter professional training in NYBG's School of Horticulture. Shortly after graduation in 1981, he went to work in the "Prop."

Interestingly, Keith points out, several of his fellow students were also trained as nurses before taking up horticulture full time. Observing his skilled hands at work, and his obvious care for living things, one begins to see parallels between the two. Nursing has also taught him the value of maintaining strict hygienic conditions when handling plants, in order to keep common orchid diseases under control.

Keith Lloyd sees his job as a twofold one: helping to preserve tropical species whose rainforest habitats are being destroyed, and sharing their beauty with visitors to NYBG. Under his curatorship, the Garden's orchid collection is becoming recognized as a leading one in the field.

A dramatic welcome to the Museum Building is provided by the Thekla E. Johnson Rotunda and Orchid Terrarium, a gift from Walter Johnson in memory of his wife. (Allen Rokach/NYBG)

Plant Hunters Extraordinary

"For a fairyland it was—the most wonderful that the imagination of man could conceive."

So wrote Sir Arthur Conan Doyle in his 1912 classic *The Lost World*, a quest for the land of El Dorado in which his fictional heroes wound up on a remote plateau in Venezuelan Guayana inhabited by wild ape men and ferocious dinosaurs. But it was left to quite a different explorer, a New York Botanical Garden scientist named Bassett Maguire, to discover the truth about this fairyland, which turned out in many ways to be more astonishing than fiction. His accomplishments, in turn, reveal much about a vital aspect of the Garden—its extensive plant-collecting and scientific programs—of which the general public is almost totally unaware.

As a boy, Maguire's favorite books were *The Lost World* and other jungle tales like *Green Mansions*, William Henry Hudson's popular 1916 novel, and as he grew up he was determined to become a botanist. In late 1953, already a veteran of nearly a decade of expeditions for NYBG, including many to Guayana, he led a small team of scientists to the highest and largest of the mountains that Doyle had fancifully described from earlier reports. It was a tablelike, 250-square-mile mesa or *tepui* rising to an altitude of nearly 10,000 feet above the Amazonian rainforest, which stretched in a trackless green carpet for hundreds of miles on all sides. Because the peak was often invisible in the clouds that swirled around it, he was to name it *Cerro de la Neblina*—"Mountain of the Mists."

Maguire first saw his goal during an earlier expedition in April of that year. He and his wife Celia, who often accompanied him on field trips, had climbed partway up another mountain near the Rio Negro when he stopped with an exclamation and pointed to the horizon, where they made out three enormous *tepuis* through a break

◄ *Shod with special tree grippers and armed with a machete, NYBG curator Scott Mori collects plant specimens in French Guiana.* (Alan Bolten)

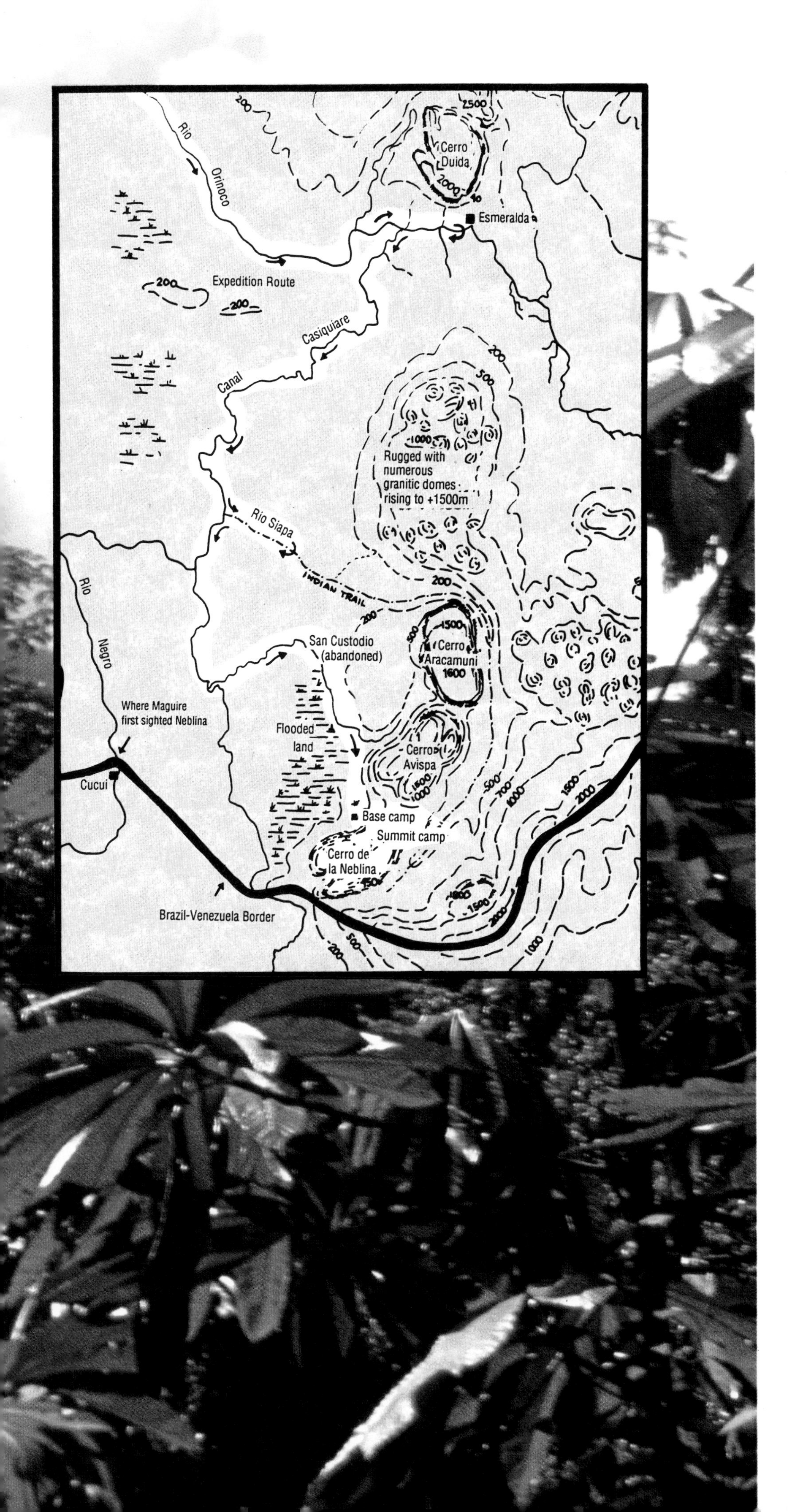

A map of the first Neblina expedition shows the party's route by boat up the Orinoco River and tributaries, then overland through the rainforest for the assault on the peak.

The expedition's target, rising from the rainforest in a swirl of clouds: Cerro de la Neblina—"Mountain of the Mists." (James Luteyn)

Neblina's explorer, Bassett Maguire, stands on the summit above the mountain's mile-deep canyon. (Celia Maguire)

in the rain. The mountains were not recorded on any map, but it is likely that they were the same ones glimpsed a century earlier by European botanist-explorers, including Richard Spruce and the legendary Robert Hermann Schomburgk, discoverer of the giant Victoria water lily. "Clearly, another expedition had to be undertaken!" Maguire wrote in a later account. He returned home to prepare for the major effort that would be required to reach his prize.

For the better part of six months he set about raising the funds required for all NYBG expeditions, then as now. He obtained major support from the National Science Foundation, and additional money from the Davella Mills Foundation and donations of material from Parke, Davis & Co., Sears, Roebuck and other companies. He assembled nearly five tons of equipment and supplies for his party, which was to involve two younger NYBG scientists, John Wurdack, a companion on earlier expeditions, and George Bunting, plus a crew of river boatmen, porters, trail cutters, hunters and others who would be hired when they arrived on the scene.

Each item was checked carefully by Maguire himself, to make sure they had everything they needed for their work, and survival, far from civilization. As recorded in his notes, these included: "Plant collecting, drying and storing equipment to the amount of 1,000 pounds; tents, tarpaulins and camp equipment, 1,000 pounds; axes, machetes, rope and other mountain-climbing gear, trade goods, clothing and sleeping equipment for the men, 1,500 pounds; medical supplies, cameras and photographic equipment, 500 pounds; and foodstuffs to last an expedition of some 25 people for three or four months, 6,000 pounds." The "trade goods," brought along to exchange for services, included cloth, sewing needles, mirrors and combs.

The explorers left New York in early October, aboard a Gulf Oil Company ship whose owners obligingly provided free passage for the men and their supplies (Ma-

Loaded to the brim, a dugout canoe carries botanists, crewmen and plant specimens on an expedition in Venezuela's highlands. (Wayt Thomas)

In rugged terrain, a helicopter and suspended "birdcage" transport NYBG botanists to otherwise inaccessible heights and depths. (William R. Buck)

guire, who knew the presidents of several major oil and shipping companies, always managed to get a ride to the field and back, thus stretching what were usually limited expedition funds). Arriving in Caracas, Maguire persuaded the American ambassador to lend him the embassy's plane and pilot for an 1,100-mile scouting flight over the target area and back, despite heavy cloud cover that made it difficult to see and dangerous to fly. In a film Maguire made of the flight, it is clear that the explorers were headed straight into a huge rock escarpment when a break in the clouds narrowly saved their lives.

In November, the party reached the remote village of San Fernando de Atabapo, a jumping-off place for previous NYBG expeditions, where they assembled a flotilla of large motorized dugouts and smaller canoes. (On his first venture into the area some years before, Maguire had been welcomed with unexpected festivity. The reason, it turned out, was that they were fascinated by his magnificent red beard. It recalled the ruddy whiskers of Schomburgk, who was still regarded as a miracle man in the legends of local tribes.)

As they voyaged upstream into smaller tributaries, which the explorers noted were the color of mahogany from decaying vegetation, they encountered 5-foot electric eels that could kill a man, as well as 18-foot anacondas and assorted venomous snakes. To conserve their staples, the party's hunters killed tapirs, peccaries and armadillos for fresh meat.

When food ran short, the visitors could fall back on various local delicacies—including giant earthworms up to five feet long, which were sliced and stewed with yams, and large spiders that when cooked tasted very much like hard-shelled crabs.

In emergencies, Maguire found, he could also get along on a diet of palm buds and bird soup.

Eventually the river deteriorated into a maze of narrow channels where the explorers' way was often blocked by huge vines and fallen trees five feet thick, taking them 10 days to progress only a dozen miles. Finally they left their craft and set out into the jungle on foot. Ropes, ladders, and hastily built pole bridges were used to get men and supplies up embankments and across rushing streams. A total of five staging camps had to be built and stocked along the way in order to relay supplies.

By December 22, the party had pitched its Camp III at the base of a steep talus slope formed by rock falling off the mountain. Wurdack and Zambo, the trail chief, were sent to scout possible routes up the slope and the cliffs above it, which rose a sheer 2,000 feet to the top.

"At five o'clock in the afternoon," Maguire wrote, "Zambo came in, wet and exhausted, and just as darkness set in we heard Wurdack's hail from the trail above. Both had labored prodigiously, only to cut two blind-alley trails—one reaching, heartbreakingly, to a bare hundred feet from the summit. Plants they brought back increased our excitement, and were prophetic of the extraordinary things we were to find. . . ."

The next morning Maguire, Wurdack and Zambo set out on a new line, but by two o'clock found themselves just below the escarpment at the end of a third dead end. It was plain the ascent had to be made farther west, so they established yet another intermediate camp in that direction for the final assault. In getting to it through the forest growing on the slope, Maguire observed that many of the trees were "giants 8 or 10 feet in girth and 200 feet high, raising their crowns above the ceiling." He added glumly, "Steady, heavy rains hampered advance, prevented our obtaining views from treetops, and kept us sodden, cold and miserable."

Forty-eight days after leaving San Fernando, Maguire and his colleagues reached the top of Neblina. It was December 31, New Year's Eve.

"For the first time in days a warming sun broke over us," Maguire recalled gratefully. "To the west and north, back across the jungle in which we had so long been immersed, dense banks of clouds reached toward the horizon, but before us lay an awe-inspiring panorama. Low [scrub] forest covered a gently sloping valley, which a mile and a half to the southeast dropped into an enormous chasm stretching to the north and south as far as we could see."

The chasm was Neblina's Cañon Grande, a rift 25 miles long and nearly a mile deep splitting the huge plateau in half—and often raging with torrents fed by the region's rainfall of 200 inches a year. "One of the world's greatest canyons," Maguire marveled, "rivaling any of the grand canyons of the United States."

To celebrate New Year's Day, the party found a flat area on which to pitch the tents of Camp V, stocking them with supplies brought up laboriously over the escarpment trail. For the next 24 days, their total stay on the mountain, Maguire reported that

◀

Insect-devouring pitcher plants, Heliamphora tatei, *growing in giant clusters on Neblina's summit.* (Joseph Beitel)

On Neblina, Maguire found tall, artichoke-like plants that he named Neblinaria celiae *after the mountain and his wife. NYBG's William Buck and others later hypothesized that the species evolved in response to lightning-set fires. Its thick, spongy bark provides good insulation even when charred, and the tender buds, held high above flames, are protected by rosettes of rainwater-collecting leaves.* (Joseph Beitel)

"collecting continued vigorously." Though the scientists encountered no dinosaurs or ape men, they were amazed by a world of which most botanists can only dream.

Among their finds were strange-looking clusters of giant pitcher plants that grew 10 feet high and bore as many as 30 skyward-pointing cups. Each cup held a quart of sticky rainwater in its base to trap and digest insects—an adaptation to lack of other nutrients, which were constantly leached away by torrential rains falling on the mountaintop's poor soil. Maguire named the new species *Heliamphora tatei* in honor of a friend, George Tate of the American Museum of Natural History, who had led previous expeditions to the region. A similar adaptation was found in bromeliads that grew huge rosettes of leaves designed to catch the rain—and in whose central pools lived small bladderworts that digested tiny organisms in order to survive.

Maguire's expedition also discovered beautiful terrestrial orchids thriving in wet peat bogs, and silvery lichens more commonly associated with the Arctic tundra. Particularly arresting were fields of weird-looking plants with thick, twisted stems, bearing what looked like open artichokes on top and producing lovely, camellia-like blooms. Identifying this as an entirely new genus, Maguire named it *Neblinaria* after the mountain, and added the species name *celiae* for his wife.

(Since some NYBG botanists spend much of their time in the field, their spouses often go along, sharing the chores. "These aren't gentle, 'fun' excursions," Celia Maguire recalls, "but hard, grueling work, gathering specimens by day and heating them in presses after dark, getting up to check the stoves all night long." In fact, not a few spouses are NYBG scientists themselves, and frequently add to an expedition their own areas of expertise.)

All in all, Neblina proved a botanical bonanza: the party brought back thousands of samples of different kinds of plants, many of which they had never seen before.

Maguire later noted: "Even from a cursory study of the specimens collected, it would seem that on the summit of Cerro de la Neblina there is a higher rate of endemism [plants unique to a region] than is now known to exist on any other tabular mountain of Guayana. Certainly more than 50 percent of the species discovered on the extraordinarily beautiful mountain are new to science."

While that figure may have been a little high, later studies have confirmed that many of Neblina's plant species are indeed unique, probably existing nowhere else on earth. The reason lies in the evolutionary history of Neblina and other *tepuis,* which are remnants of a vast plateau that dates back hundreds of millions of years to the time when Africa and South America were once joined as a single continent. The sandstone of which the formations consist is relatively impervious to the torrential rains that gradually eroded away the softer rock and earth around them, eventually leaving them as towering islands in a jungle sea.

Each *tepui,* isolated from the others and the countryside at large, became a sort of landlocked Galapagos whose flora and fauna developed along somewhat different lines, influenced only by the occasional wandering bird that might drop an alien seed on the mountaintop, or by the rare animal or insect that ventured up its steep, windswept slopes. In its way each of these mountains, indeed, has become a "lost world"—a laboratory in the clouds where scientists can test ideas of how life evolves.

Having established Neblina as his scientific "turf," Maguire returned on subsequent expeditions in the late 1950s and 1960s to continue his work. Interest in the mountain continued to grow, so much so that in the early 1980s Cerro de la Neblina became the focus of major international investigations that included as many as 125 biologists from various institutions eager to examine not only the mountain's plants but its birds, bats, insects and other animal life. Appointed to coordinate the work of

Mycologist Roy Halling gets close to the ground to photograph a tiny species of Galeropsis. (Barbara Thiers)

the botanists, which included 27 from the United States—11 of them from NYBG—was a younger Garden curator, James Luteyn. Also along was Brian Boom, newly appointed as Curator of Guayana Highland Botany to carry on Maguire's research.

NYBG fern specialist Joseph Beitel, in a Yankees baseball cap, displays prize specimens gathered on the mountaintop. (James Luteyn)

NYBG fern specialist Joseph Beitel, who collected more than 200 species on Neblina, including a handful of new ones, described the experience as "*Caramba!* botany—every time you saw something else, you shouted *Caramba!*" Mycologist Roy Halling, lowering his lanky frame close to the ground, discovered button-sized mushrooms that were totally unfamiliar—and scenery so otherworldly that he remembers standing on the plateau one day and muttering to himself over and over, "Oh my God." Wiliam Buck, a specialist in mosses, liked the rare pleasure of being able to work with researchers in different fields: "To be on a trip with all those other scientists was like being a kid again. You could run around and catch frogs and bugs, only now you had someone to tell you about them."

The later expeditions to Neblina were made somewhat easier by helicopters, which were hired to ferry the scientists from their base camps to the top, bypassing Maguire's arduous struggles up the cliffs. Though old-line explorers might sniff at such modern luxuries—and some did—the participants still sampled many of the same hazards that faced their predecessors, and have always confronted scientists in the field. The most constant one on the mountain was weather—windy, cold, unpredictable, and often very wet. Luteyn remembers setting up camp one night in what seemed like a nicely sheltered montane valley, only to wake up in a cloudburst with water lapping over the top of his cot. On another occasion during the expedition, one scientist came across a patch of what looked like nice, plump blueberries and triumphantly brought back a bucketful to share with his colleagues for breakfast. Those who feasted on the treat had something to say about the donor's generosity. They spent the rest of the day huddled in their tents with nausea and cramps.

During their travels, Garden botanists have experienced almost every other kind of surprise. (As the great Linnaeus pointed out more than two centuries ago in his *Critica Botanica:* "Good God! When I consider the melancholy fate of so many of [botany's] votaries, I am tempted to ask whether men are in their right minds who so desperately risk everything else through their love of collecting plants.") NYBG adventurers have fallen off log bridges into streams; broken bones when thrown out of Jeeps; been threatened by trigger-happy border guards; and have been bitten by snakes and assaulted by malaria, yellow fever and other tropical diseases. More than one has played host to botfly larvae, which, once implanted under the skin, live off human blood and tissue until they become so large and painful they may have to be surgically removed.

Wayt Thomas, an NYBG specialist in sedges who has traveled to many parts of tropical America, still hasn't gotten used to the most relentless occupational hazard—bugs.

"I don't mind snakes so much," says Thomas. "Even the poisonous ones usually get out of the way. But the insects are something else. On some trips I've averaged four or five wasp stings a day, and jungle wasps can be pretty fierce. The worst, though, are probably the ants, particularly solitary ants an inch long that manage to bite and sting you at the same time—a neat trick using both ends. One species of ant in Brazil is known as the 'twenty-four', because that's at least how many hours you feel the effects."

In his office in a sixth-floor garret atop the Garden's old Museum Building, Thomas shows an inquiring visitor what he calls his collection of "Oh, My!" insects, carefully gathered on his travels and mounted in rows. Among them is a wasp as big as a hummingbird, with a nasty curved stinger like a fishhook and a wingspan nearly five inches across.

"When you hear that character droning through the jungle like a bomber," he observes, "you duck."

Once a botanist finds a plant worthy of preserving, whatever hardships have been involved, his job has barely begun. Indeed, plant collecting, far from being glamorous, involves so much tedious manual labor that botanists sometimes refer to themselves as "hay balers."

Time spent collecting is usually matched by an equal amount in camp, where the collector spreads out his samples—as many as a dozen of each plant to provide duplicates for study and sharing with other specialists—then presses each carefully

Wayt Thomas, used to insects as an occupational hazard, grins and bears a swarm of stingless bees. (Courtesy of Wayt Thomas)

On Neblina and other high tepuis, explorers often contend with cold, torrential rains. Here a NYBG party ponders its flooded camp. (James Luteyn)

Herbarium worker Steve Crisafulli checks plant specimens in the cool storage room. (Allen Rokach)

between sheets of newspaper for drying. Specimens must be labeled with exact dates and locations, notes on soil, temperature and moisture, and other data including frequency and association with other plants. A live specimen must also be photographed in color where it grows, to record its true appearance and surroundings (the color of dried specimens, and bulkier or fragile ones that must be preserved in alcohol, soon fades). If a plant looks particularly promising, live cuttings of new growth are made, wrapped in moist tissue and placed in a plastic bag to determine if the species can be successfully grown in the Garden's greenhouses back home.

The field-pressed specimens are taken to base camp, where they are stacked between pieces of heavy blotting paper, interleaved with corrugated aluminum spacers to allow air circulation, then heated in a drying press over a portable kerosene or gas stove until the water in the leaves and stems has been largely removed. Finally, batches of the dried specimens are carefully boxed and taken to the nearest major airport, where they are shipped back to New York by air freight.

When they are received at NYBG's herbarium—one of the world's largest with some 5.2 million specimens on file—the boxes go as soon as possible into a freezer, where they are kept at 0 degrees Fahrenheit for 48 hours to kill any insects, eggs or other organisms that may have come along for the ride. The samples, still in sheets of newspaper, are then removed from the packing boxes and stacked ceiling-high in a cool storage room maintained at 52 degrees. There they remain until they can be examined in greater detail by the returning botanist or other specialists.

A scientist first attempts to classify a new specimen by comparing it with drawings and descriptions of similar species in the Garden's library, checking monographs (papers that provide descriptions and keys to known species of a given family) or floras (treatments of all members of a group of plants in a given geographic region).

◄

Back at camp after a day in the field, Scott Mori presses his collected specimens in sheets of newspaper for drying and shipping. (Carol Gracie)

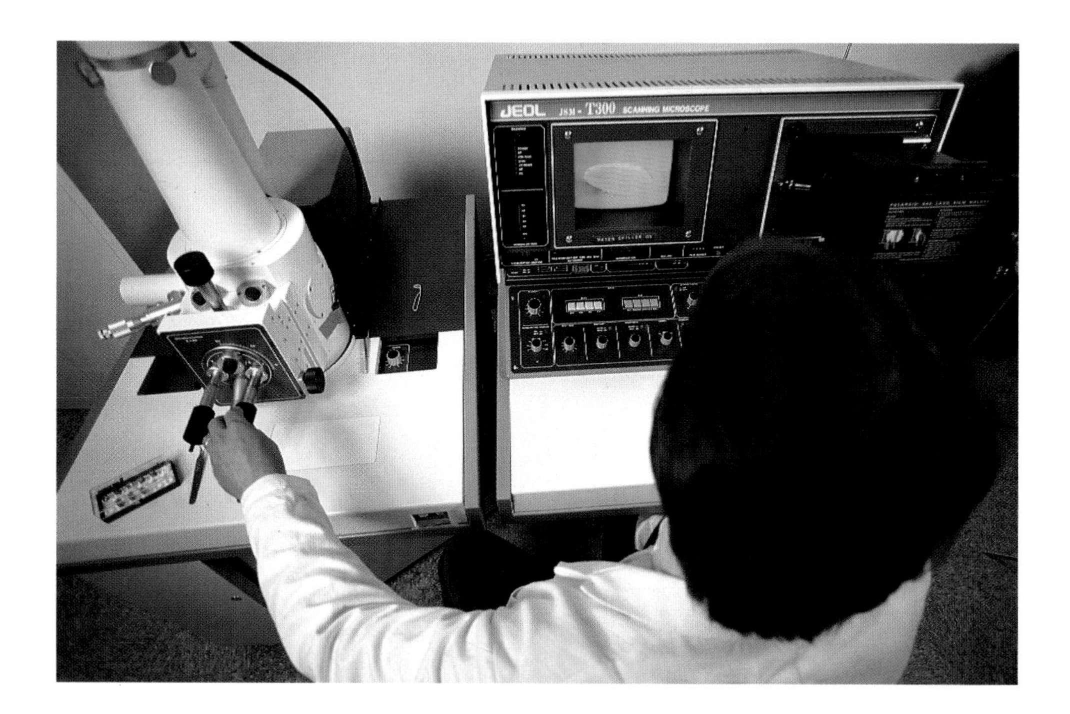

In NYBG's laboratories, Donald Black uses a scanning electron microscope to study plant cells, pollen grains and fungus spores. (Allen Rokach/NYBG)

A plant that cannot be identified in any of these sources is taken to the herbarium, where it is compared with identified specimens of the same family and genus to try to find a match (a process wryly referred to as the "herbarium crawl").

To aid the process of identification, a botanist may use a dissecting microscope, which enables him to view parts of flower, leaf or stem in detail. If needed, he can also have specimens prepared and examined by a skilled technician in NYBG's Charles B. Harding Laboratory Building, using a scanning electron microscope with a power of 10,000 or more to magnify tiny cells in even greater detail. He can also ask for help from the laboratory's plant chemist or anatomist to settle questions about the plant's functions or pedigree.

If all this research confirms that a new plant species has been discovered, the specimen is given the exalted status of a "type" specimen, to be used as a reference for all future studies of the plant. Some 125,000 type specimens assembled from various sources are carefully filed on the Museum Building's fourth floor, in the only area of the present herbarium that has full air-conditioning.

The botanist then prepares a formal description of the species, with a Latin preface and other precise terms that other botanists understand (see box), giving it a genus name, comparable to a person's surname, and a species name, comparable to a first or given name except that it follows rather than precedes the genus name. Accompanying the description is a drawing by a botanical artist, an exacting and delicate art form in itself. The description and drawing are then published in a journal read by other botanists.

The discoverer of a new plant, or fresh facts about an old one, doesn't have to look far to make his findings known to fellow botanists—the Garden's scientific publications program, begun in 1896, has grown into the most extensive of any botanical

Botanese

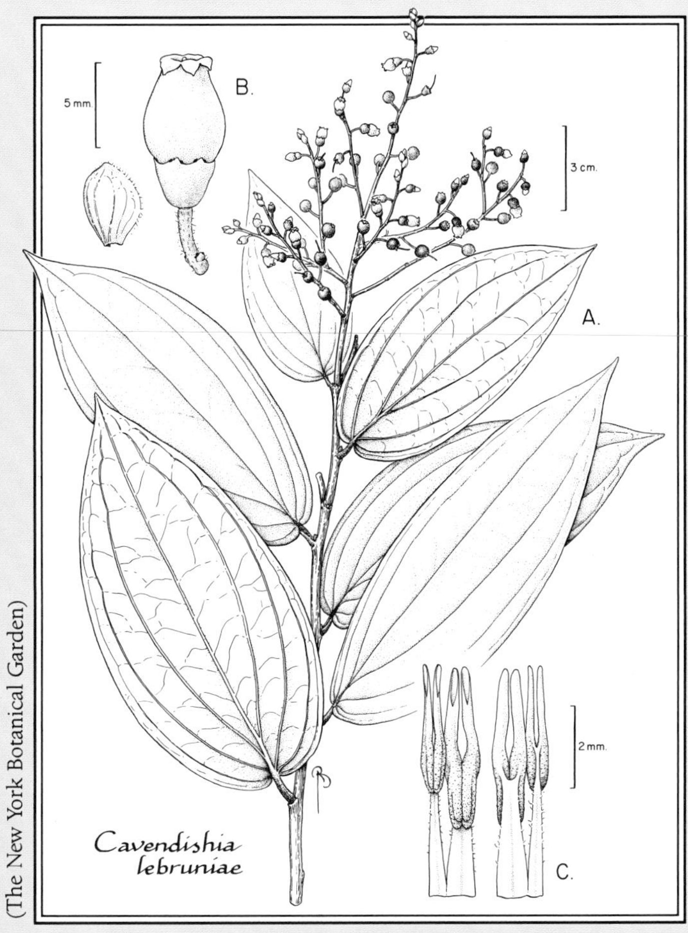

Cavendishia lebroniae Luteyn, sp. nov. (Fig. 1)

Quoad inflorescentiae habitum ad C. martii *et* C. divaricatam *accedit, sed ab eis glandulis calycis pubescentia et floribus differt.*

Terrestrial shrub arching to 2.5 m from mostly single stems. *Mature stems* terete, glabrous, brown. . . . *Immature stems* and twigs of new growth bluntly angled, striate, densely short pilose, light brown, glaucous. *Petiole* terete, rugose, 5–15 × 2–4 mm, densely pilose, dark reddish-brown. *Leaves* coriaceous, elliptic or ovate-elliptic, 10–27 × 3.5–10 cm, basally rounded to slightly subcordate, apically short acuminate, entire, sparsely short pilose along the nerves on both surfaces, also with scattered glandular fimbriae on the lamina beneath. . . . *Corolla* urceolate, 9–10 mm long and 8 mm in diam. . . .

On a plant-hunting trip in Ecuador, NYBG botanist James Luteyn, a specialist in the blueberry family, and his associate, Maria Lebrón, were clambering up a gravelly slope when they spotted an interesting shrub. It turned out to be a new species, which Jim named *lebroniae* in honor of her. Says Maria with a smile: "He told me the real reason was that the flowers were short and pudgy, and I didn't like that too much."

In publishing his findings in *Brittonia,* one of the Garden's professional journals, edited by senior curator Noel Holmgren, Luteyn described his new species in language that a layman might find impenetrable but that botanists around the world can immediately understand.

Following the genus to which it was assigned (named in 1836 after an early botanical patron, William Cavendish, Sixth Duke of Devonshire) appears Maria's surname, properly Latinized. This reflects an accepted practice in which a new plant species, or genus, is often named after a person, though the name can reflect a physical attribute of the plant itself (e.g., *megabracteatum*, meaning very large bracts) or the place where it was found (e.g., *pennsylvanica*, for Pennsylvania). After the name of the discoverer is an abbreviation for "species novum," signaling that this is a species new to science.

Next, a Latin "diagnosis" tells botanists that the general shape and arrangement of the flowers resemble two other species in the genus, but that *lebroniae* differs in the glands of the calyx, or flower base, as well as the flowers themselves and nearby hairs. In a lengthier English description, excerpted briefly here, are other precise botanical terms—terete (round), glabrous (without hairs), pilose (softly hairy), glaucous (covered with a whitish bloom), rugose (wrinkled), coriaceous (thick and leathery), subcordate (partly heart-shaped), acuminate (pointed) and so on.

Buried in the description is the term "Corolla urceolate." It means that the petals are urn-shaped—a comparison for which Maria hasn't quite forgiven the author yet.

(Maria Lebrón)

garden in the United States and any herbarium in the world. Presently under the direction of Maria Lebrón, a plant ecologist by training, the program includes nine scholarly journals edited by members of the Garden's scientific staff and mailed to scientists in 90 countries. Chief among these are *The Botanical Review* (established in 1935), a quarterly survey of important developments in the botanical sciences; *Brittonia* (started in 1931 and named in honor of the Garden's founder and first director), which publishes research papers in systematic botany from around the world; *Economic Botany* (1947), the official publication of the Society for Economic Botany, and *Advances in Economic Botany* (1983), both specializing in relationships between plants and people; and *Mycologia* (1909), a bimonthly dealing with fungi and lichens and the official publication of the Mycological Society of America. Others include *North American Flora,* begun in 1900 to provide descriptions of all plants growing in North and Central America and the West Indies, and *Flora Neotropica,* which has published accounts of plants in the New World tropics since 1976.

In addition to its journals, NYBG has published more than 30 books of general as well as scientific interest, including a two-volume set on the native orchids of the United States and Canada, others on rhododendrons, rare plant conservation, botanical illustration and a floristic inventory of tropical countries. Recent books include a continuing series on *Intermountain Flora* and a second edition of Arthur Cronquist's classic *Evolution and Classification of Flowering Plants* (see box). Even better known to the general public is the monumental *Wildflowers of the United States,* the first definitive work of its kind, published between 1966 and 1975 by McGraw-Hill. Written by botanist Harold William Rickett in collaboration with other staff members under the editorship of William Steere, it consists of six volumes of two to three books each on

Dorothy D'Alisera glues dried specimens in the herbarium's mounting room. (Allen Rokach/NYBG)

the Northeast, the Southeast, the Southwest, Texas, the Northwest and the Central Mountains and Plains. All in all, NYBG believes, it has produced more guides to the flora of the Americas than any other institution in the world.

While scientists prepare reports on their expeditions, one of six technicians in the herbarium's mounting room takes each of their specimens and, arranging leaves and stems gently, affixes them to a stiff sheet of acid-free paper with water-based glue. The mounted specimens are weighted down to allow the glue to dry, then sewn with thread around the roots, stem and larger leaves to make sure they stay firmly in place. Loose seeds and plant parts are carefully saved in a small envelope attached to the sheet; bulkier fruits and mushrooms are stored in separate, labeled boxes, and some

(The New York Botanical Garden)

A Taxonomic Titan

YOU CAN TELL WHEN he is around. First comes the voice, a rich basso rendering a selection from *Rigoletto* or *Figaro*, which echoes among the metal cabinets in the herbarium's high-ceilinged halls. Then around a corner comes the man himself, a towering figure whose bushy white eyebrows flare upward as if prepared for flight. It is Arthur Cronquist, all six feet six inches of him. NYBG's senior scientist, he is a giant in the science of plant taxonomy and the author of texts used round the world.

Cronquist, who grew up in California, Oregon and Idaho, describes himself as a "nature boy" who liked getting outdoors to hike or climb, the reason he first majored in forestry at Utah State. In his senior year, he decided to go into taxonomy, or systematics, instead "because that was what I was spending most of my time on anyway." After taking his Ph.D. at the University of Minnesota, he joined the Garden in 1943. Except for a hiatus from 1946 to 1952, he has been there ever since, serving as editor of the *Botanical Review* for the last two decades and author of major volumes on the flora of the United States.

Even as a student, Cronquist had been aware of inadequacies in the system of classification devised by the German Adolph Engler—a revision of Linnaeus' original that botanists had followed for almost a century, but that omitted newer ideas about the evolution of plants. "People generally agreed that the Engler system had to be recast," says Cronquist. "But they also said it was just too big a job for any one person. I got tired of hearing that story, so in the mid-1950s I decided to give it a try."

It took a bit of work. In 1957 Cronquist published a paper "staking a claim" to the subject, then started to assemble a book, *The Evolution and Classification of Flowering Plants,* which was finally published in 1968. Encouraged by its success, he added to and refined his own system in *An Integrated System of Classification of Flowering Plants,* which came out 13 years later in 1981. "A monumental tour de force," intoned the journal *BioScience*. A reviewer in the magazine *Science* found the tome—at 1,262 pages—"rather too cumbersome for convenience."

Rupert Barneby, a respected senior curator at NYBG, believes the book has a different kind of weight: "History will consider Cronquist's system one of the great benchmarks, a state-of-the art report of this century's understanding of plant taxonomy. There isn't a botanical laboratory in the world that doesn't have a copy—usually a well-worn one."

Most of his fellow scientists agree. In 1985 Cronquist received the American Society of Plant Taxonomists' prestigious Asa Gray Medal (which has been given only five times—most recently to Rupert Barneby). The next year he was awarded Britain's highest botanical honor, the Linnean Medal, placing him in the company of history's great botanists.

Art Cronquist has his own explanation: "I'm something of an ivory tower type, and I'm too ornery to work for anyone else."

cacti and algae are sealed in jars of alcohol. (Once in a while a technician may complain of odd rashes on her hands. Even though some plants have been frozen and desiccated for a year or more, they can still have enough powerful chemicals left in them to irritate the skin.)

The mounting crew produces more than 80,000 new specimen sheets a year, a painstaking job but one that often produces results of delicate beauty. The sheets are filed in long rows of gray steel cabinets that take up most of the Museum Building. The specimens are arranged in a systematic sequence intended to show the plants' degree of similarity and their evolutionary relationship, from more primitive ferns

Senior curator Rupert Barneby examines a mounted specimen through a microscope. (The New York Botanical Garden)

and gymnosperms on the sixth floor to the most advanced species of the flowering composite family on the first. Duplicate sheets of specimens are often made so the first can be returned to the country of origin; other duplicates may be sent to botanical gardens in the United States or abroad, either on loan for study or in exchange for desirable specimens from their collections. The Garden's herbarium has the most active loan program of any in the world. In a recent year 67,729 specimens were lent to 57 herbaria in 34 states and to 95 herbaria in 31 foreign countries; 23,523 specimens were borrowed from other herbaria for study at NYBG.

To a casual visitor walking past the endless rows of metal cabinets—more than 2,000 of them—the herbarium might appear to be nothing more than a graveyard of dead plants. But to botanist Patricia Holmgren, its director, and to NYBG's scientific staff of more than 70, it is the soul of the Garden—a treasure house of science and an active international center for comparative research. Indeed, to a dedicated botanist, a fine plant specimen is as cherished as a Shakespeare quarto is to a bibliophile, a composition of genetic information as intricate as a Bach fugue.

In addition to plants collected by the Garden's own explorers, the herbarium boasts more than a score of collections acquired from colleges, institutions and individuals. These range from the 400,000 specimens deposited on permanent loan by Columbia University in 1899—the start of the NYBG collection—up to such recent prizes as 40,000 fungi acquired from the University of Massachusetts in 1989. The oldest specimens on file are ones from Captain James Cook's eighteenth-century voyages to the Pacific, acquired on exchange from the British Museum, and others from Lewis and Clark's and John C. Fremont's early nineteenth-century explorations of the American West, which came as part of Columbia's John Torrey collection.

Though NYBG's collection is diverse and cosmopolitan, its greatest strength is in plants of the Americas, the focus of the Garden's curators since the start. Any scientist who attempts a serious study of New World flora, in fact, must consult NYBG's herbarium—a major reason why some 250 botanists from other institutions and nations arrive to work in the collection each year, some staying for many months at a time.

The urge to collect is universal; to the botanist it is essential. And though dried herbarium specimens may be invaluable for study, they are no substitute for observing live species in their native habitats. Since Per Rydberg's first trip to Montana in 1897, NYBG scientists have completed close to 900 expeditions on all seven continents, studying both individual plants and entire ecosystems. In the beginning they largely worked close to home in the northeastern United States, gradually moving west to the mountain regions, north to Alaska and south to Jamaica, Puerto Rico and other Caribbean islands—where the amazing diversity of tropical species beckoned them even farther afield.

Patricia Holmgren, director of the herbarium, would rather be out looking for live plants. Here she poses with a specimen of elkweed (Frasera speciosa) *on a Utah plateau.* (Noel Holmgren)

An Amazonian Romance

A South American legend tells of a native chief's daughter whose unrequited love for the moon finally caused her to drown herself in a lake. Filled with compassion, the moon transformed her body into a beautiful flower, which opened its petals to welcome the lunar deity each night. The Indian girl might have been the inspiration for the fabulous *Victoria amazonica,* which was named instead for England's queen.

This giant among water lilies—a favorite Victorian picture shows one of its huge floating pads supporting a small child—was discovered and named by Robert Schomburgk in 1837, during an expedition for the Royal Geographical Society of Great Britain. Plants raised from its seeds soon became major attractions at Kew and other botanical gardens around the world, including NYBG, in whose Conservatory pools specimens are displayed today. But it was not until the 1970s that a Garden scientist, Ghillean Prance, discovered what actually happens to the flower at night.

Prance and a colleague, entomologist Jorge Arias, observed that although in the tropics *Victoria amazonica* blooms throughout the year, each flower lives for only two days. On the evening of its first day, when the white, magnolia-like blossom unfolds to a width of a foot or more, it emits a heady scent described as a mixture of pineapple and butterscotch.

Whatever the odor conjures for humans, it appeals greatly to scarab beetles, which enter and are trapped when the flower closes for the night. In a cozy interior chamber, which the plant maintains at a temperature warmer than the surrounding air, the beetles feast on starches that the plant produces specifically for their tastes; safe from predators, they also while away the hours with newfound beetle mates. All this is part of the water lily's plan: in the process the beetles deposit pollen picked up on their bodies from other Victorias, thus fertilizing their host.

The petals remain closed for 24 hours, and when they open the next evening their color has changed from pure white to a blushing cerise. As the beetles exit, they brush past the blossom's anthers, picking up fresh pollen. They carry the grains on their bodies as they proceed to the next opening Victoria, which they will dutifully fertilize in the course of another blissful night.

For Ghillean Prance, shown here with his subject, the story has an added twist. A tall, bearded Scotsman, and a veteran plant hunter who served the Garden for 25 years (most recently as senior vice-president for science and director of the Institute of Economic Botany), he has returned home to serve England's present queen. In 1988 Prance was appointed director of Britain's Royal Botanic Gardens, Kew—completing a circle that began when the Brittons founded NYBG on Kew's model a century ago.

For the past half century the Garden's primary concentration has been on South America, whose rainforests and mountains contain the greatest diversity of plant life on earth. The majority of the world's flora and fauna has evolved in the tropics, but only a small fraction of them have actually been identified. Moreover, the sheltering forests and understory plants are constantly being destroyed for farming, cattle raising, logging and developments like hydroelectric dams. At the rate things are disappearing, scientists guess, tens of thousands of species could become extinct in the next few decades, before they can be discovered and analyzed. For the Garden's scientists, it is a race against time.

Associate scientist Dennis Stevenson, for example, travels all over tropical America as well as Africa, assessing ways to preserve and propagate the ancient cycads, rated as endangered everywhere they occur. Curator Scott Mori, a specialist in the Brazil nut family, investigates the ways that trees and animals interact to perpetuate themselves in the central Amazon region of Brazil. Douglas Daly, Steven Churchill, James Luteyn, Barabara Thiers and others conduct detailed inventories of the rich and threatened floras of Colombia, Peru and Ecuador. Brian Boom, Laurence Dorr, John Mickel and Lisa Barnett study the forest trees, ferns and medicinally useful plants of Venezuela. Andrew Henderson works in Haiti in hopes of preserving one of the world's rarest palms. Richard Howard, the Garden's vice-president for botanical science, takes time out from administrative duties to catalog the flora of the Lesser Antilles and to study ways of replanting areas denuded by strip mining in the Dominican Republic, Jamaica and Hawaii.

Along with all this activity in the tropics, North America remains an ongoing priority. Clark Rogerson, senior curator of mycology, hikes mountain areas from the Great Smokies to Utah's Wasatch Range to catalog the fungi of the United States. During 25 years of research in the intermountain region between the Sierra Nevadas and the Rockies, senior curator Noel Holmgren alone has discovered and named 28 new species and seven varieties of flowering plants.

The organizing structure behind all these efforts is taxonomy, the study of natural diversity in which plants and animals are classified into related groups based on factors common to each. These groups start with the broadest one, a phylum, more commonly called a division in botany, and branch down through class, order, family, genus and species.

Plant taxonomists today more often describe their work as systematic botany, a regimen that gives scientists the world over common methods and a universal language with which to explore similarities and relationships among plants, and thus to investigate the processes of evolution itself. Systematic botany is the keystone of all the plant sciences, and it provides critical support to the other biological sciences as well. An herbarium, in turn, is a keystone of systematics; without reference to its collections, neither basic nor applied research can go very far.

Systematists at NYBG, like their counterparts at other institutions around the world, study the ecosystem in which a given plant grows in order to determine its place in that system. They examine the plant in detail in an attempt to decipher its genetic composition and heredity. They seek to understand how mutation and natural selection have, through the ages, shaped the plant into its present form. They study the plant's means of reproduction, and they count the number of each species in a given area to measure fluctuations in its population. They examine the chemical minifactory that each plant must have to survive, hoping to find secrets in those chemicals that may prove useful to man.

Information from all these investigations is valuable not only to botanists. It provides a scientific basis for rating rare and endangered species, enabling conservationists to find out what is threatened and where, thus to establish priorities in setting aside nature reserves. Corporations and government agencies call on NYBG scientists and their data to identify commercially valuable, toxic or illegal plants, and to help prepare environmental impact statements and land management plans. Projects have ranged from serving as expert witnesses for law enforcement agencies concerned with marijuana to the identification and control of species poisonous to sheep and cattle on the western range.

Knowledge gained through plant hunting and analysis can also prove vital to ordinary people, sometimes in unexpected ways. Thousands of requests for plant identification or advice come to the Garden each year from police departments, hospitals and poison control centers, as well as from the public at large. By consulting reports in NYBG's library, the Garden's Plant Information Service is able to handle most of the questions on the telephone. Sometimes, however, even the librarians are stumped, and have to refer a knotty problem back to the herbarium or to members of the scientific staff.

Howard Irwin, a botanist and former president of NYBG, vividly remembers one such case:

"Early one morning in July my telephone jangled impatiently," he recalls. "The caller was an official of the Board of Health of the City of New York. 'I have some seeds that need to be identified today,' he said. 'Can you help me?' 'Of course,' I told him. 'Bring them around.' "

An hour later Irwin and the official were examining the contents of a little sack that had been confiscated from a tourist at New York's Kennedy Airport. It was a collection of beautiful brown, yellow, black and multicolored seeds. One immediately caught Irwin's eye.

"It was the spectacular jequirity, or crab's eye, a glossy, pea-sized seed of Chinese red tipped with intense black. As we knew, one jequirity seed, if chewed and swallowed, will release enough phytotoxin to kill an adult. The beans, nevertheless, are used by West Indians in rosaries, necklaces and bracelets."

On Irwin's advice, customs agents were instructed to watch carefully for further importations of the deadly seed. How many lives of curious children were saved by this timely action will never be known.

Map by Georg Brewer Botanical drawings by Bobbi Angell

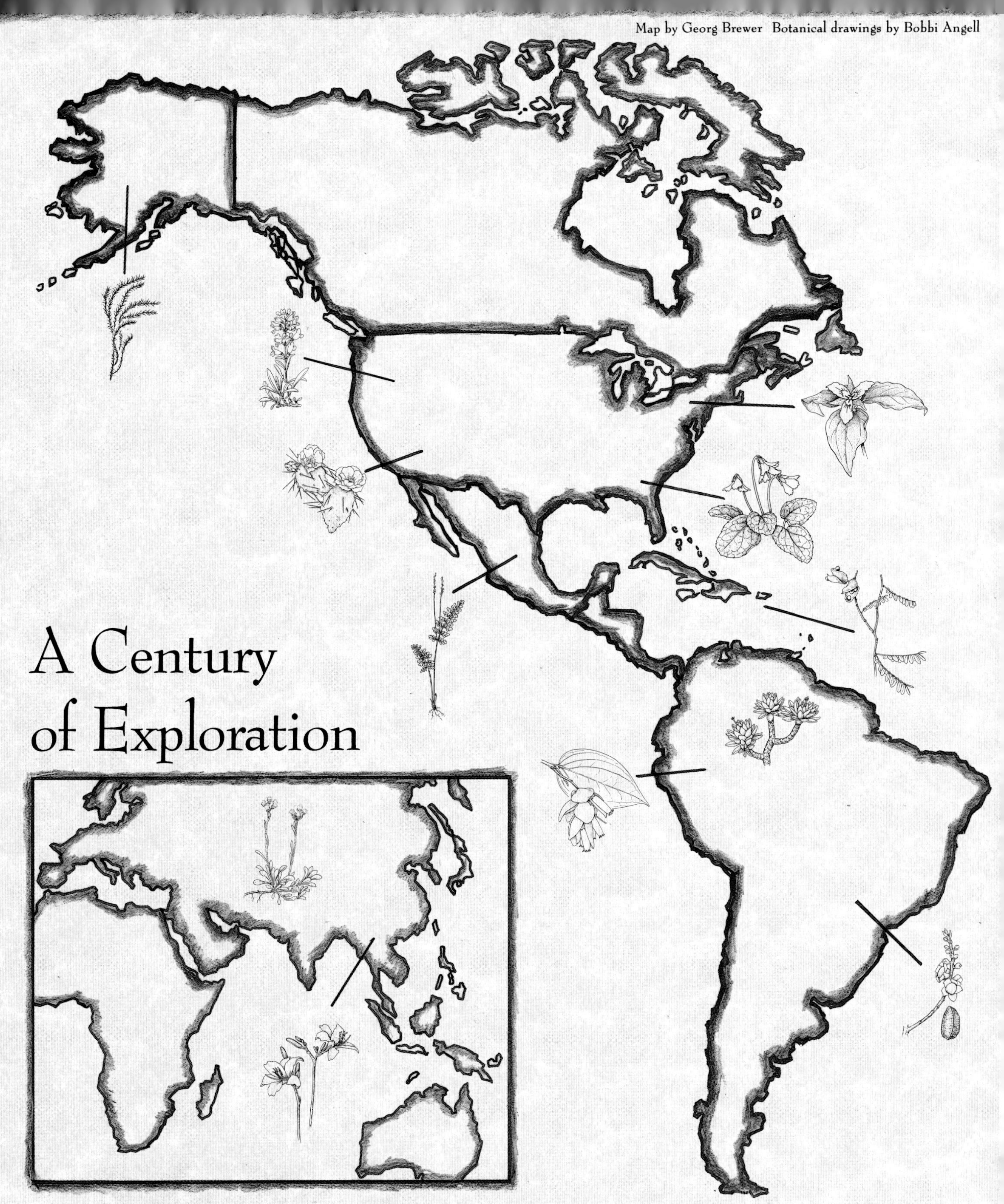

A Century of Exploration

SINCE ITS FIRST FORAY TO MONTANA IN 1897, THE NEW YORK BOTANICAL GARDEN HAS conducted nearly 900 research expeditions around the world. From frigid tundra to steamy rainforest, Garden scientists have collected hundreds of thousands of plant specimens for study, published comprehensive inventories of regional flora, sought out new sources of food, medicine and fuel. For the last half century, their efforts have increasingly concentrated on tropical Latin America, probably the greatest mother lode of plant life on earth. For details, see the following two pages.

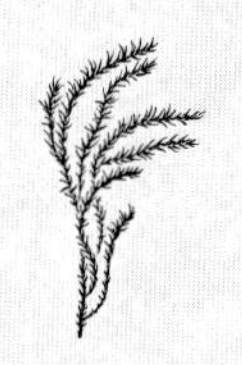

Andreaea nivalis

Alaska and the North

Much knowledge of the world's northern flora can be traced to the research of William Steere, a former NYBG president, who made extensive collections in the tundra of Alaska, Canada, Greenland and Iceland. His work in bryology (mosses and liverworts) remains so influential that the Garden maintains a fund for visiting scientists to study his specimens.

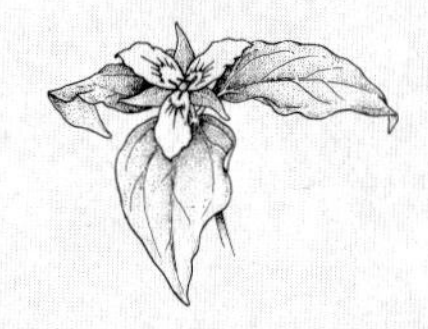

Trillium undulatum

Northeastern United States

NYBG's first major survey resulted in the 1896 publication of the *Illustrated Flora of the Northern United States, Canada and the British Possessions,* written by the Garden's director, Nathaniel Lord Britton, and Addison Brown, and since revised in manual form by Henry Gleason and Arthur Cronquist. The Garden maintains a current interest in the region's flora with Cronquist's new edition of the *Manual of Vascular Plants of Northeastern U.S. and Adjacent Canada.* Nationwide, the Garden's most famous popular work is its 14-volume *Wildflowers of the United States.*

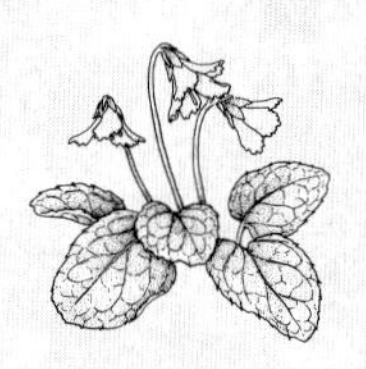

Shortia galacifolia

Southeastern United States

Botanist John Small began collecting here before 1900, when the region was still largely rural. His massive *Flora of the Southeastern States,* first published in 1903, remains basic to current studies, which have been updated by NYBG contributions to later publications.

Penstemon compactus

Western United States

Following Per Rydberg's expedition to Montana, and publication of his *Flora of the Rocky Mountains and Adjacent Plains* in 1917, Garden botanists have made many studies of the American West. Notable is Arthur Cronquist's co-authorship (with C. L. Hitchcock of the University of Washington) of a five-volume work and field manual on plants of the Pacific Northwest. A major series, still under way, is *Intermountain Flora*, begun in 1972 by Cronquist, Noel Holmgren, Patricia Holmgren, Arthur Holmgren and James Reveal.

Opuntia

Southwestern United States

After years of field work in the early part of the century, Nathaniel Lord Britton and Joseph Rose published their monumental four-volume work *The Cactaceae,* the first comprehensive survey of the cactus family.

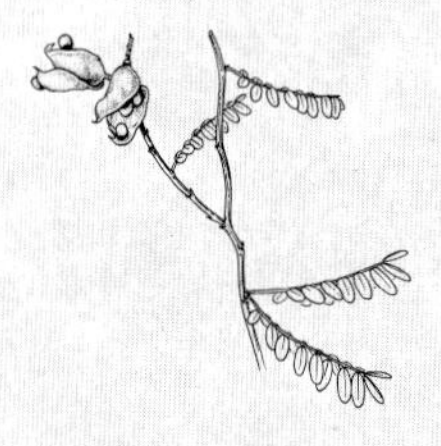

Abrus precatorius

The Caribbean

Early expeditions by Britton and other staff members produced pioneering studies of the flora of Bermuda, the Bahamas, the Virgin Islands and Puerto Rico. Current efforts include surveys of West Indian mosses by William Buck and of the ferns of Hispaniola by John Mickel.

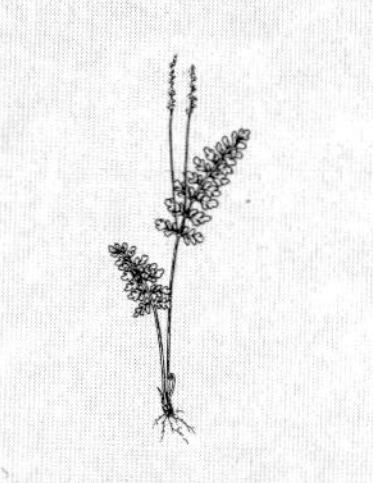

Anemia hirsuta

Central America and Mexico

A long-standing interest in the region continues with the studies of John Mickel and Joseph Beitel, who have produced a major volume on the ferns of Oaxaca, and with the work of NYBG's Institute of Economic Botany, whose director, Michael Balick, has collaborated with local scientists and native healers to research the practical uses of native palms and medicinal plants.

Neblinaria celiae

Venezuela

In exploring Neblina and other table mountains of the Guayana Highland, Bassett Maguire and a succession of other NYBG researchers have uncovered many species found nowhere else in the world, including three new families of flowering plants.

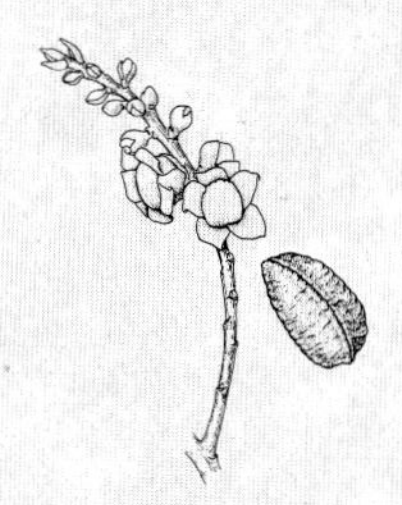

Bertholletia excelsa

Brazil

Some 140 expeditions have been mounted to different parts of South America's largest nation, and more are planned. NYBG's interest in the Amazon region stemmed from explorations made during World War II by botanists seeking sources of much-needed rubber and quinine. A major focus today is on cooperative efforts with Brazilian scientists to catalog the threatened flora of the Amazonian basin, the severely decimated Atlantic coastal forest and the Planalto or high plains.

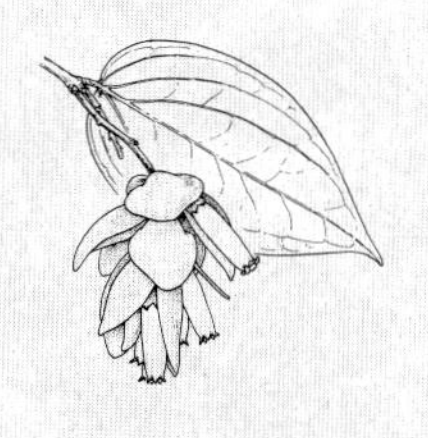

Cavendishia antioquiensis

Western South America

Studies are under way to define the plant life of Colombia, Bolivia, Ecuador and Peru, which is among the richest in the world. In these and other areas of tropical America, the Institute of Economic Botany works with native peoples to develop practical alternatives to deforestation, and to collect species for testing their potential against cancer and AIDS.

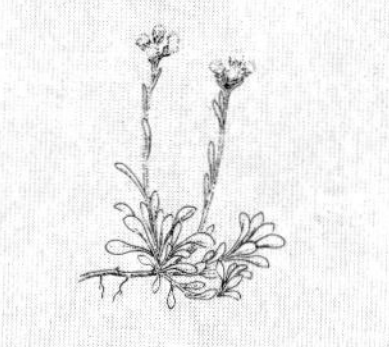

Antennaria dioica

Europe and the USSR.

NYBG botanists frequently travel to Europe and the Soviet Union to consult major herbarium collections and to work in the field with fellow scientists, who also come to the United States. Cooperative projects have led to the introduction of new varieties of ornamental plants at the Garden and its Mary Flagler Cary Arboretum in Millbrook, New York.

Hemerocallis

Asia and Australia

New research projects are under way in Indonesia to study native agriculture and to develop productive uses of tropical forests. Garden botanists also cooperate with their counterparts in Australia to enhance knowledge of important plant groups like the pea family (Leguminosae).

A Garden of Learning

What does a plant need to Stay Alive? It needs: Soil, Water and Light.

—YOUNG NYBG GARDENER

ON A BRISK DECEMBER MORNING, 24 THIRD GRADERS FROM NEW YORK P.S. 71 troop into NYBG's "Greenschool" on the Conservatory's lower level, shed jackets and mittens and squirm into seats.

"Today we're going to talk about the tropical rainforest," announces Greenschool teacher Judith Fitzgerald, eyeing a couple of fidgeting boys. "Does anyone know what that is?"

Twenty hands shoot up. Judith points to a bright-eyed eight-year-old, who ventures that, "It's a warm place where it rains a lot."

"Very good, Nicole." The teacher beams, then invites her audience to join her on an expedition to probe the jungle's darkest mysteries.

Wearing a T-shirt with an ocelot on it ("Does anyone know what 'endangered' means?"), Judith weaves a story of plants and animals that live in the forest's layers, explaining their roles and giving children large colored cut-outs of each to place on a tackboard in front.

Toward the end of the session everyone gets a surprise treat, made from three of the plants they have been hearing about—chunks of banana dipped in warm chocolate and shredded coconut. Revitalized by plant power, the kids troop out, two by two, to examine towering banana plants, coconut palms and other wonders of the tropics in the greenhouses upstairs.

Each school year some 40,000 children, ages 5 to 18, come to the Garden for such two-hour minicourses and other programs; among the most popular, besides "The

◄ *A "Greenschool" class, touring the Conservatory, makes some discoveries about aquatic plants.* (Allen Rokach/NYBG)

Tropical Rainforest," are "The Plant World," "The Desert," "Foods from Grasses" and "Plants We Drink." More than 600 youngsters take part in weekend workshops and NYBG's Children's Garden, where they learn how to grow vegetables and flowers in well-tended plots. In their own schoolrooms, still others listen raptly as visiting Garden speakers relate adventures in the Amazon, or sow seeds in lighted "Grow Boxes" and observe the marvels of growth that result.

And that's just for kids. For some 11,000 of their elders each year, NYBG offers courses in everything from tree pruning to taxonomy, plus lectures and special events. For students pursuing plant-related careers, the Garden's own School of Horticulture and other programs supply training for professional jobs.

Since its first public lecture in 1895,* The New York Botanical Garden has grown into a sort of horticultural/botanical university where there is something for everyone, regardless of experience or age. Its curriculum is the largest of its kind in the United

(Muriel Weinerman/NYBG)

A Garden for Kids

FROM SPRING THROUGH fall, one of the busiest—and noisiest—places at NYBG is the Children's Garden, where groups of youngsters aged 3 to 16 swarm around plots they have planted with tomatoes, lettuce, peppers, radishes, parsley and basil, set off by towering sunflowers. The budding gardeners, who come from a wide range of ethnic and economic backgrounds, number 350 to 400 a year, about evenly distributed among girls and boys.

*"The Rise and Progress of the Royal Botanic Gardens, Kew, England," given by Kew's assistant director, Daniel Morris.

"Kids like physical exertion, and gardening offers plenty of that," says Catherine Eberbach, who runs the garden as well as children's weekend workshops and family programs for NYBG. "They love to dig. They love to get dirty, and if they can get wet and muddy, too, so much the better."

Even youngsters who are disbelievers at first soon get into the swing. "Often a child will look up and say 'This is *fun!,'* as if he or she couldn't quite get over it," Catherine continues. "Children learn by having fun, by playing, by *doing* things—cultivating the earth, planting seeds, examining what comes up, finding out what a tomato really tastes like when it's ripe."

Since children usually enjoy the company of their peers, groups of five-year-olds, six-year-olds and on up through older grades are teamed together and assigned to their own garden plots. The youngest work in pairs, on beds 4 feet square that are raised to distinguish them from paths; older ones cultivate plots 4 feet wide and about 16 feet long.

"We try to simplify things, to make it easy, so no one will get frustrated or discouraged," explains Catherine. "We want them to succeed. If they can do something successfully, they'll want to come back for more."

Children can enroll in spring, summer or fall sessions, or all three if they like. The spring program, which runs for 11 Saturday periods of two-and-a-half-hours each, starts in early April, when older children start to plan their gardens and younger ones learn the difference between a shovel and a hoe (the littlest use kid-size tools, donated by the Smith & Hawken Co.). In the nine-week summer session, beginning in early July, youngsters come for two hours two mornings a week. The first day each week is devoted to gardening, the second to nature-related activities like forest walks, insect studies, flower arranging and poetry writing.

(Allen Rokach/NYBG)

In early September the children invite parents and friends to a harvest celebration, where they make tortillas and corn-husk dolls from corn they have grown, compete in baking contests and watermelon hunts and otherwise enjoy the fruits of their toil. The fall session, which consists of two-and-a-half-hour periods on Saturdays, continues for another six weeks as participants grow their own late crops, make compost, turn over the soil and learn how to put a garden to rest for winter.

Generally the children take home what they have grown, but some groups have donated produce from special gardens to a local shelter and soup kitchen instead.

"At first I wasn't sure how they'd react, giving away something they had worked so long to grow," says Catherine Eberbach. "But the kids were smiling when the van came to pick up their vegetables. They were really proud to give their food to someone who needed it."

States, with several hundred programs offered by an education department that numbers 30 employees and 300 part-time instructors, interns and volunteers, including outside experts and specialists from the Garden's horticultural and scientific staffs. For more than a decade, educators from other botanical gardens, nature centers, museums and zoos have been coming to annual NYBG seminars to learn how to run successful programs of their own.

"We try to reach the broadest possible audience," says John Reed, vice president for education, director of the Garden's library and a 25-year veteran of NYBG. "A lot of people, particularly those who grow up in cities like New York, don't realize how much the world depends on plants. Youngsters who come to our classes, for example, don't know that the cereal they ate for breakfast actually came from corn or wheat or oats that once grew in a field. They think watermelons grow underwater, and that spaghetti comes from a far-off place called New Jersey. Many children, and even some adults, are afraid to go into our 40-acre forest because they think bears or wolves are waiting for them in there. You might say we have our work cut out for us."

"A surprising number of schools teach almost nothing about plants," adds Rosemary Kern, NYBG's director of education. "Many children have never seen a forest or pond, and a visit to the Garden may be the first step in changing a child's perception of the natural world. In our adult programs, too, we offer people an education they can't get anywhere else. If the subject is plants, we teach it, from botanical illustration to horticultural therapy."

An outsider, particularly one who recalls old *New Yorker* cartoons, might picture the typical NYBG student as a well-upholstered matron in a large floppy hat, a Westchester garden clubber who likes to dabble in flower arranging while catching up with the girls. If the stereotype ever had any validity, Kern points out, it doesn't anymore: "Today our students cut across age groups and income levels, and they're not just trying to fill idle hours. Quite a few of our students work in high-pressure fields such as finance or advertising, and for them gardening or landscaping is a change of pace, a way of relieving stress. Some have found it so rewarding as a hobby that they have left the fast track to go into full-time horticultural careers."

The Garden's adult program reflects a veritable explosion of interest in gardening, which in recent years, according to national surveys, has grown into the most popular outdoor activity in the United States. The range of courses also illustrates how much plants have become intertwined with different aspects of contemporary life, from ecology to art. Some courses consist of one-day introductions to subjects; others extend over weeks or months. Students can take any course that appeals to them; those with special interest in one area can proceed through a prescribed curriculum and earn a NYBG certificate in that field.

Some of the most popular courses, not unexpectedly, come under the general heading of gardening, which includes garden history and contemporary garden design; gardening with perennials, wildflowers, vegetables, herbs, orchids and other kinds of plants; and such practical matters as composting and mulching, soils, pest and disease control, propagation, pruning and general garden maintenance.

Among the best attended is "Fundamentals of Gardening," taught by Ralph Snodsmith, well-known author, host of radio's "Garden Hotline" and an NYBG teacher for more than 20 years. Snodsmith's goal is getting beginners to understand "the difference between roots and shoots," and his method encourages them to do their own "detective work" in order to discover basic principles in cultivating plants. Like other

A budding florist practices techniques in a flower arranging class. (Allen Rokach/ NYBG)

good teachers, he also knows the value of humor: "If I can get my students to laugh a little in class, they're more likely to remember what I tell them."

The same attitude is evident in another teacher, Stephen Tomecek, who makes a course called "Soil Science for Gardeners" come alive. Students in his first class enjoyed it so much they gave him a nickname that he has been proudly wearing on a badge ever since. It reads "Dr. Dirt."

Another popular area is landscape design, which attracts an increasing number of students who wish to become more knowledgable about gardens in general, or to acquire skills for work in the nursery business or establish their own practices in garden design. For professionals and would-be professionals, a program in commercial horticulture provides training for jobs in retail garden centers, nurseries, greenhouses, parks, botanical gardens and landscaping firms. Individuals interested in full- or part-time employment as floral designers can take an intensive five-week program in commercial flower arranging, which delves into everything from basic centerpieces to wedding and funeral design. One measure of the program's success is that it attracts students from as far away as Japan.

An increasingly important specialty is served by NYBG's program in horticultural therapy, in which students learn how to use live plants and plant crafts to help rehabilitate patients in hospitals and nursing homes, psychiatric facilities, prisons and other institutions where the therapeutic value of working with plants is recognized.

NYBG's certificate program in botany covers plant physiology, taxonomy, morphology, genetics, ecology and field botany. Another program in botanical art and illustration, designed for both amateurs and career artists, ranges from basic flower drawing to advanced styles and printing techniques. Noncredit courses are also of-

Participants in the botanical art program take advantage of good weather, and ample subject matter, to sketch outdoors. (Allen Rokach/NYBG)

fered in botanical crafts, whose students cook herbal dishes and make their own holiday ornaments, wreaths, basketry and decorative Easter eggs.

To reach students elsewhere in the metropolitan area, NYBG sponsors some of its courses in cooperation with other institutions, including the Horticultural Society of New York in midtown Manhattan, the New Canaan Nature Center in Connecticut and the Department of Parks in Ridgewood, New Jersey. Such courses, plus original programs in ecology, are also offered at NYBG's Mary Flagler Cary Arboretum north of the city in Millbrook, New York (see Chapter 7).

In addition to its general offerings, the Garden sponsors several programs leading to academic degrees. It is particularly proud of its state-licensed School of Horticulture, which was established in 1932 to alleviate a shortage of skilled gardeners for private estates in the New York area and was reorganized in the 1970s to offer broader training for professional careers. Most students take the intensive, 21-month program full time, spending half their hours in class learning botany, gardening and landscape design, the other half gaining practical experience by working with the staff in NYBG's own gardens, Conservatory and Propagation Range. Not a few graduates, who earn a diploma in horticulture, wind up in jobs in botanical gardens, including NYBG itself.

On other levels, cooperative programs with the City University of New York lead to associate and baccalaureate degrees in botany, horticulture and landscape design. In its efforts to train more scientists needed in today's increasingly vital fields of systematics and environmental conservation, the Garden's science department provides valuable experience for graduate students from many universities who work in NYBG's laboratories and herbarium and accompany senior botanists on research expeditions in the field. The Garden also offers international fellowships to scientists from developing nations, who spend up to a year receiving practical training in economic botany at NYBG.

For its scientists, its students and many others, the Garden's library is an indispensable resource. Begun in 1899 with the permanent loan of 5,000 volumes from Columbia College's library, and supplemented by an aggressive program of acquisitions, it is now the largest and most actively used of its kind in the United States, sharing its body of primary reference works with both visitors and other institutions around the world by means of interlibrary loan. Its holdings—which total close to a million items, including books, journals, manuscripts, artwork, archives and photographs—encompass botany, horticulture, landscape design, garden history and other plant-related subjects, with particular strength in systematic botany, plant geography, ecology and the history of botanical science.

In addition to scientific and popular texts, the library's collection includes a number of rare books and manuscripts which are housed in a special room upstairs and frequently used for the important historical information and illustrations they contain. The library's archives are the repository for major research and historical collec-

(The New York Botanical Garden)

A Medieval Treasure

ONE OF THE VALUED holdings in NYBG's Library is *Circa instans*, an elegantly illuminated twelfth-century text that experts have called the most important herbal of postclassical times, a botanical equivalent of the Dead Sea Scrolls.

Compiled around A.D. 1140 by Matthaeus Platearius, a magister (teacher) at the School of Salerno, the first school of medicine established in Europe, the text was written on vellum in medieval Latin by a skilled calligrapher. Lacking a title, as is normal with medieval manuscripts, the book is identified by the first two words of its prologue, which reads "About the present matter of medicinal simples. . . . "

Circa's authorship was deduced from a reference in the text to a painful ailment that was treated with a medicinal plant called "strucium" (known to later botanists as *Saponaria officinalis*, soapwort or Bouncing Bet), and from the statement "By virtue of this remedy I, Platearius, was made free." Included are descriptions of this and hundreds of other plant-derived drugs, as well as instructions in their preparation and use. It reflects the School of Salerno's determination to cut through a muddle that surrounded medicine in those days, to reduce a welter of common names into a codified group that physicians could use, and to provide precise descriptions of botanical raw materials to help them avoid adulteration and fraud.

Up to then herbalists had relied slavishly on Dioscorides' *De Materia Medica* and Pliny's *Historia Naturalis*, both written in the first century A.D.

By breaking the mold, *Circa instans* influenced botanical thinking for the next three centuries, and thus is recognized as a key work in the development of modern botany. The book was obtained in the 1970s from the library of Dr. Emil Starkenstein, who had been a professor of pharmacology at the German University in Prague before World War II.

(The New York Botanical Garden)

tions, including the records of the Torrey Botanical Club, the Society for Economic Botany, the American Society of Plant Taxonomists, the Mycological Society of America, the Council on Botanical and Horticultural Libraries and the Organization for Flora Neotropica. The library is custodian of the comprehensive files compiled for the definitive bibliography *Taxonomic Literature,* which is heavily used by researchers, including NYBG botanists, who must trace the development of knowledge about specific plants back to its origins.

Also available to scholars are the manuscripts and correspondence of noted botanists like John Torrey and Dr. David Hosack (Chapter 2); a collection of the works and correspondence of Charles Darwin; and the notes of inventor Thomas Alva Edison, who carried out research in a NYBG laboratory in an attempt to derive rubber from goldenrod. Among other valued holdings are the works of outstanding botanical illustrators like Mary Eaton, Anne Ophelia Dowden and Pierre Joseph Redouté. There are also hundreds of old nursery and seed-company catalogs that document the development of American gardening and horticultural tastes in colorful detail. A special collection consists of the architectural drawings and records of Lord & Burnham, one of America's oldest greenhouse firms and the designer and fabricator of NYBG's Conservatory.

John Reed, vice-president for education and director of the library. (Allen Rokach)

"Early on, Nathaniel Lord Britton saw the importance of establishing a major botanical library in America," says John Reed, "and he put tremendous energy into acquiring the needed literature, a tradition we have tried to carry on. Our collections now represent more than 75 percent of the world's literature on plant systematics and floristics. That's pretty remarkable in the botanical world considering the fact that our library is less than a hundred years old.

"And these publications don't just sit on the shelves gathering dust," Reed adds. "They are consulted all the time, because systematic botany builds upon the published work of the past, and the first description of a given plant may go back three or four hundred years. Also, if one of our botanists is planning an expedition to a certain area of Mexico, for example, he will come to the library to find out everything he can from the accounts of earlier botanists who explored the same or similar areas in the past—what they collected, and where. A particular species he wants to study may be recorded as growing only on top of one mountain, and he needs to know how to find it, providing it's still there, of course."

The library's card catalog and small reading room are in constant use by some 8,000 individuals a year, including not only the Garden's staff and students, but by visiting botanists from other countries; writers doing research for articles and books; professional artists seeking to render flowers in accurate detail; representatives of pharmaceutical firms and other businesses looking into potentially useful plants. Librarians at NYBG also answer more than 4,000 requests a year for information over the telephone or by letter with correspondents as far away as Denmark, Brazil, India and the Philippines.

Though many queries are handled speedily, some take hours of digging, and others require a little patience and tact. "You always know when a school class gets an assignment," adds research librarian Bernadette Callery. "All of a sudden twenty letters arrive, saying 'Dear Miss Botanical Garden: I am writing about botany. Please send me everything you know.' You may laugh, but you can't dismiss those inquiries. Nor can you ignore the eighth grader who writes in asking what one has to do to become an ecologist, because he really wants to know." Reference librarian Lothian Lynas recalls one lad who started using the library in his early teens to pursue a passion for insectivorous plants—and is now one of the leading authorities on the subject in metropolitan New York.

To share its information with its clients—and with other institutions, which in return may be able to supply material NYBG lacks—the library is linked with more than 9,000 libraries in North America and Europe through the Online Computer Library Center (OCLC), an international bibliographic data base that indicates which libraries have which books. In the 1990s, the Garden and sister institutions hope to develop this into an integrated computerized system to improve information-sharing and coordinate the acquisition of new works. Such a system would also eliminate cumbersome card catalogs, allow many more points of access to a given subject and

(Allen Rokach/NYBG)

"Miss Information"

WHAT DOES POISON IVY LOOK LIKE—AND CAN I CATCH IT FROM MY HUSband? My cat is eating my houseplants. What should I do?

Answering such questions was all in a day's work for Elizabeth Cornelia Hall, the gentle, lively mastermind of NYBG's Plant Information Service until her death in 1989 at the age of 91. For 52 years "Miss Hall" (no one addressed her by her first name) served the Garden faithfully—30 years as head librarian, 22 more as a volunteer after her "retirement" in 1967. Her knowledge of plants became legendary, as did her joy in sharing that knowledge with others. Says former NYBG president James Hester, "She was undoubtedly the most beloved person in the New York horticultural world."

As a girl, Elizabeth Hall wanted to become a doctor, but when she found that Harvard Medical School did not allow women, she took a degree in chemistry at Radcliffe instead. Attrracted by the "farmerette" movement of the 1920s, she learned how to raise vegetables and prune trees in the School of Outdoor Life for Women at Philadelphia's Temple University, then became one of America's first horticultural therapists by introducing plants to patients at the Pennsylvania Hospital for Mental and Nervous Diseases.

In the late 1930s, while working at the New York Horticultural Society, she was persuaded to come to the Garden by T. H. Everett, whom she later helped in compiling his 14-year, 10-volume opus *The New York Botanical Garden Illustrated Encyclopedia of Horticulture*. She always knew where to find answers to questions, and she always called him "Mr. Everett." He used to call her "Miss Information," at which both would laugh. Everett attributed her longevity to a steady diet of double martinis for breakfast, but quickly added, "Miss Hall is the best horticultural librarian in the world today!"

Mr. Everett and Miss Hall liked to run the Plant Information Service together, fielding telephone queries with zest. They grieved sympathetically with owners of dead or dying houseplants, but when a woman asked how she could kill an unwanted tree without her husband's knowledge, Miss Hall replied sweetly, "Perhaps you could kill your husband first."

She remembered one day when two officers from New York's Sing Sing prison came into her office with some weeds in hand: "They said that during exercise periods inmates would pick them up in the yard, boil them and get high in their cells! I was able to tell the officers it was jimsonweed, which contains a powerful alkaloid. They were very appreciative, but they insisted on my getting a doctor to sign a statement to that effect."

On another occasion it was pets, not prisoners, who were going wild. Miss Hall listened to a tale of dogs digging crazily in a bed of unidentified plants, consulted an old herbal, and came up with the answer—valerian, from whose odorous roots modern tranquilizers, as well as rat poisons, are derived.

Moreover, she informed the caller with some satisfaction, it wasn't the Pied Piper's flute that drew all those rats out of Hamelin to drown in the river. It was the valerian, clever fellow, that he carried in his sack.

make the resources of all member collections immediately visible on computer screens at NYBG and around the world.

Across the hall from the library, and working closely with it, is the Garden's Plant Information Service, which answers queries from the general public on everything from lawn maintenance to poisonous plants. Long watched over by the incomparable Miss Hall (see box), the free service is now run by Dora Galitzki, a specialist in woody plants, who with volunteer help responds to some 6,000 telephone calls, letters and visits a year. For answers she doesn't know offhand, she consults NYBG's own 10-volume *Illustrated Encyclopedia of Horticulture* or other books on her copious reference shelf. If she's really stuck, she calls on specific Garden librarians or scientists familiar with the field in question.

"You never know what people are going to ask," says Dora. "One day a lady called to say that her fig tree was singing at night; I told her perhaps a cricket or cicada had flown in through an open window and taken up residence, and she said thanks, she'd look into it. Another woman announced that her parakeet was dead and wanted to know which one of her houseplants was responsible. I had a little trouble handling that."

Perhaps the oddest query came from a man who announced that he had been invited to a Halloween party and had to have his costume ready in a couple of weeks. "No goblins for him," Dora recalls. "He was determined to go as the U.S. Open golf tournament, if you can believe it, and he wanted to know how to grow grass on his clothes! I thought for a moment, then suggested that perhaps he could make a muddy mixture of grass seed, soil and water, plaster it on an old tweed jacket he didn't care about, and keep the whole thing moist and warm—and to use rye grass because it would grow faster. Of course I never heard how things worked out, but you might say that's par for the course."

For the general public, NYBG also provides a lively agenda of visitor activities, including "family days" with tours and hands-on demonstrations; theme weekends celebrating the peak bloom of daffodils, azaleas and other seasonal displays; guided walks through the Forest, Rock Garden, Native Plant Garden and other areas on the grounds. Large-scale public symposia are held at the Garden and other locations on topics ranging from landscaping with perennials and wildflower cultivation to Japanese gardens and contemporary landscape design.

Learning extends into the field, both near and far. Frequent day trips are led by knowledgeable guides to outstanding natural areas and to public and private gardens around metropolitan New York. Foreign tours and cruises, with accompanying experts, are offered each summer to such destinations as the gardens of France and the British Isles.

For the more adventurous, NYBG scientists lead small groups to their special areas of expertise, including the Amazon, the Galapagos Islands, Mexico and other tropical sites. Travelers not only have the chance to see exotic species and cultures first-hand, but on some trips are able to learn in greater depth by helping with actual plant collection and research.

One such expedition, which has been repeated for several years, is a two-week exploration of the remote rainforests of French Guiana. Led by NYBG curator Scott Mori and his wife Carol Gracie, also trained as a botanist, the trip is limited to eight paying participants, who use as their base of operations the small homestead of a French family that is a four-mile hike from the nearest landing strip. During the day, the volunteers and their guides fan out to collect specimens for an ongoing study of the area's flora; in the process they learn about the region's complex ecology and observe the roles of insects, birds and other animals in pollinating plants and dispersing seeds.

Another expedition has been a 10-day trek through the forests and pine savannahs of Belize. Michael Balick, director of the Garden's Institute of Economic Botany, leads a small group of volunteers in collecting and processing plants that will be analyzed by the National Cancer Institute for their potential in treating cancer and AIDS (Chapter 7).

Closer to home, the Garden has become increasingly aware of its reponsibilities to the community immediately outside its gates. In 1988, with the help of funds from the city and private foundations, NYBG launched its Bronx Green-Up program, aimed at helping residents make their borough a more beautiful, and ultimately more stable, place in which to live and work.

The Bronx, once an area of rural farms and estates, had gradually been transformed into small, middle-class neighborhoods in which homeowners took considerable pride. In recent decades, however, many of the old families dwindled as they or their children moved away; buildings were abandoned, burned or torn down; park land was neglected and barren sidewalks replaced pleasant tree-lined streets.

Working with borough officials, the city's parks department, Operation Green Thumb and volunteer groups, NYBG established a new office and staff to coordinate efforts to "green" the Bronx. While not a wholly original idea—similar projects were under way in inner cities around the country—it was particularly appropriate for the Garden, which had long felt uncomfortable about being an oasis for higher pursuits amid surroundings that presented such vivid contrasts.

Projects have included planting flowering bulbs around run-down subway stations and along unsightly fences, as well as starting gardening programs for senior citizens, children's day care centers and schools. The most significant efforts, however, have been in encouraging neighborhood groups to start community vegetable and flower gardens on city-owned vacant lots, for which NYBG, the city and other organizations supply free advice as well as soil, plants, seeds and tools as they are made available through donations.

Starting with 20 gardens in the first year, Green-Up director Terry Keller hopes to broaden the program to include hundreds more (the Bronx, often referred to as a "bombed-out area" by residents and visitors alike, has thousands of vacant lots, so the supply is not likely to run out).

The first person to draw on the fledgling program was Sister Cecelia of the Crotona Community Coalition, who called in early spring to announce that her group had cleaned up a lot at Prospect Avenue and 181st Street. She had 80 Girl Scouts and 25 neighborhood people ready to dig in, but they didn't have any soil. Through the Rockefeller Corporation, Keller located free topsoil in New Jersey and enlisted the city's Parks Department to pick it up; she also arranged for compost, manure (from the neighboring Bronx Zoo), fencing materials, lumber for raised gardening beds, potted perennials and annuals and several hundred packets of donated seeds. By summer, Sister Cecelia's lot was lush with beans, tomatoes, lettuce, peppers, zinnias

(Allen Rokach/NYBG)

NYBG's Magic Box

TO INVOLVE A WIDE AUDIENCE OF CITY CHILDREN, NYBG'S EDUCATION Department has developed a "Grow Box" program for New York area schools. Well over a hundred of these miniature laboratories have already found their way into classrooms and day-care centers, where they help teach science to students from preschool level to 12th grade.

Not unlike light boxes used by adult indoor gardeners, the Grow Box has a welded aluminum framework 4 feet wide, 3 feet high and 2 feet deep. Hung from the top are four light fixtures, each with two fluorescent tubes, which can be lowered to provide light directly over small seedlings and raised to accommodate the plants as they grow. An automatic timer turns the lights on for the optimum period every day—12 to 14 hours to bring plants like tomatoes and peppers to maturity indoors—and can be adjusted for experiments in growth.

Vegetables, herbs and flowers are raised under the lights in individual pots, which are placed in plastic trays filled with perlite and water to provide moisture. The unit is delivered and installed complete with trays, perlite, pots, soil mix, seeds, fertilizer and labels. As part of the package, NYBG provides a three-hour training session for teachers and a curriculum manual with lesson plans.

"To children, it's almost like magic; they can sow seeds in their own classroom in the middle of winter and watch their plants grow day by day," says Director of Education Rosemary Kern. "They can also get a taste of the scientific method—research, preparation, experimentation, observation. Not least of all, it's a chance to learn something about nature, and to begin to appreciate the difference that plants make in their lives."

and marigolds. It had already been given a name by its builders—"the Garden of Happiness."

"So many in the community are affected by the gardens," says Terry Keller. "You can see it in the smiles of passersby, in the comments of those who stop to chat. People realize they *can* change their environment. One neighborhood now has five gardens within a three-block area, and is working on getting safe, unbroken sidewalks and street trees."

Green-Up participants agree. "Gardening teaches children to take care of their community," says Theresa Ocasio, who led her Girl Scout troop in transforming a rubble-strewn lot into a flourishing garden at Clinton and 182nd streets.

"It brings white, black and Hispanic people together, the old and the young," says José Lugo, who works as a volunteer in a community garden a few blocks away. "Everyone watches the garden, so no one comes to do drugs here anymore."

Another amateur gardener, Karen Washington, describes the promise of a greener city in even more eloquent terms: "There is something going on in the South Bronx that is not all crack, burned-out buildings and poverty. We are turning the desert into an oasis. We are doing something with nature and giving it back to the community."

Bronx Green-Up participants pose proudly in a community garden they developed from a vacant lot. In front are Felix Graham (left) and Bernie Saunders. Behind them, Lois Reddick and Robert Smith flank the program's director, Terry Keller. (Allen Rokach/NYBG)

Green Gold and Acid Rain

"You wish to be what you have always been, and you think you cannot go beyond what you are. I assure you that this is the best formula for suicide for a scientific institution."

At a 1980 symposium convened by President James Hester at the New York Botanical Garden, a panel of leading scientists and educators wrestled with the role of botanical gardens in a changing world—a world that seemed increasingly to threaten the very environment on which it depended for life. The proceedings went along smoothly enough until the distinguished microbiologist and author René Dubos took the podium, startling his audience with the words above.

Dubos' warning was not mere rhetoric. He urged that botanical gardens, in the interests of their own survival if nothing else, concentrate their energies in three areas of broad social consequence: (1) finding new plants useful to mankind, (2) developing ways to humanize urban areas, (3) devising methods to restore damaged ecosystems to health.

While his comments provoked some heated arguments—purists wanted to stick to the time-honored traditions of botany—the challenge led NYBG's staff and board through eight months of soul-searching debate.

The result was the formation of three new institutes under the Garden's auspices: an Institute of Economic Botany, to focus on plants that could be used for food, fuel, medicine and other human needs; an Institute of Ecosystem Studies, to study disturbances and recoveries in communities of plants, animals and microbes; and an Institute of Urban Horticulture, to develop ways to improve deteriorating city environments with plants.

A decade later, the first two of these offspring are alive and well, and making notable contributions in their fields. (Due to lack of initiative and funding, the Insti-

◀ *On a 65-foot-high scaffold, IES researchers Eileen Geagan and Jean Hubbell enclose branches of a white pine tree in clear plastic bags. The needles are exposed to varying concentrations of ozone and acid rain to determine if these pollutants accelerate the loss of nutrients from foliage.* (Ted Spiegel)

tute of Urban Horticulture is no longer an entity, though many of its functions are carried on by the Garden's horticultural and educational staff, including the community projects to "green" the Bronx described in Chapter 6).

The Institute of Ecosystem Studies focuses on problems affecting northern temperate ecosystems, and their global ramifications. Among them are the effects of air pollution on forests and lakes; the damage and recovery of natural communities from both human activities and natural disturbances such as windstorms and fires; and the dynamics of plant diseases and pests.

Headed since its inception in 1983 by Gene Likens, a leading ecologist and a discoverer of acid rain in North America, IES and its scientists work out of the Garden's "other campus," the Mary Flagler Cary Arboretum, a 2,000-acre preserve 75 miles north of New York City in Millbrook, New York.

The arboretum, unlike its parent in the Bronx, came about almost by chance, through the generosity of a single individual—a lady who loved trees. The land originally served as a summer retreat for Melbert and Mary Flagler Cary, who had acquired 14 farms near Millbrook. The granddaughter of Henry Flagler, a tycoon in Standard Oil and Florida real estate, Mrs. Cary felt great attachment for her hideaway, which she called Cannoo Hills, and in her will she specified that it be preserved.

After her death in 1967, the friends she had appointed as trustees, with the subsequent help of Edward Ames—then an officer of the Ford Foundation and now of the Mary Flagler Cary Charitable Trust—asked for proposals from a number of conservation-minded groups. They settled on one submitted by Howard Irwin, then president of the New York Botanical Garden, who outlined a program of horticulture, science and education as the most productive use of the land. And so, in 1971, the Mary Flagler Cary Arboretum was born—happily, with money from the trust to help run it.

The early administrators focused on developing an arboretum in the traditional sense. They converted a handsome 1817 brick mansion into a visitor and education center; established notable collections of pines, birches, willows and dwarf conifers from the United States and Asia; planted a fern glen with more than a hundred hardy native and foreign species; constructed roads and interpretive trails for visitors; built a greenhouse to propagate plants for the collections, and later to display tropical plants.

The 1970s saw the beginnings of educational programs for schoolchildren and adults, as well as scientific projects that included attempts to breed a disease-resistant elm. In coping with the land's burgeoning deer population—which had a dismaying habit of dining on the arboretum's prized new plantings—Cary researchers developed an ingenious slant-wire electrified fence that the deer had trouble jumping, and that has proved a model of its kind.

It was not until the mid-1980s, however, after the new Institute of Ecosystem Studies had moved into a low, modern office and laboratory building constructed earlier, that the arboretum began to fill its special niche in the scientific world. Today the IES staff has grown to some 80 people, closer to 140 in summer with the addition of research assistants and interns. Significantly, it boasts a score of Ph.D. scientists, more than many universities have in the field of ecology.

"In many ways we're unique," says Likens, who left an endowed chair at Cornell to meet the challenge of building the new institute. "Of all the botanical gardens around the world, I don't know of any that has a program like this. And we're not caught up in a university-style bureaucracy, where there can be enormous amounts of red tape. We try to find the brightest minds we can, give them the resources and freedom they need, then turn them loose to pursue the best leads in science, to capitalize on serendipity. And we don't keep our findings under wraps."

Likens cautions against the hope for easy solutions in ecology, which attempts to examine factors affecting the interactions among organisms, and among organisms and their environments: "We're dealing with some very complex problems of nature at a large scale, which become evident only by study over long periods of time. These problems are difficult to sort out and usually require more than one mind. We and other ecologists around the world are building a pyramid of knowledge. Here at IES we interact, we talk, we argue, sometimes we disagree vigorously. And every once in a while something really important pops out."

Such discoveries come slowly, and often by chance. They are made by what Likens calls "bright scientists keeping their eyes open," and they must be developed carefully to see what the initial evidence really means. An example is the institute's current leadership in the study of air pollution, which actually started back in 1963 when

IES's Cloud Water Project measures airborne pollution atop Mount Washington in New Hampshire. (Andrew Cott)

Likens was leading a study of forest ecosystems in the White Mountains of New Hampshire at a site called the Hubbard Brook Experimental Forest, near the town of West Thornton.

"We were measuring the chemistry of rain and stream water to see how forests take up and release nutrients," he recalls. "From the beginning we said to ourselves, that's curious, I wonder why the pH of the rain is so low?"

What Likens and his colleagues had discovered was acid rain. Since there were not yet enough data to put it in perspective, they continued to amass further information to bolster their suspicions, finally publishing their findings in 1972. "We settled on the term 'acid rain' for the title," says Likens. "But we found later that a scientist named R. A. Smith, observing the air in London, first referred to it as such almost 100 years before."

In little more than seven years since IES came into being, its scientists have carried out scores of studies at different sites around the country, including the now-famous Hubbard Brook, to determine more precisely how forests, streams and lakes react to air pollution and other man-made factors, as well as how they recover from logging and from natural events like windstorms and volcanic eruptions. These studies, which are published in scientific journals, are outlined in an educational report called *Discoveries in Ecology*, a summary for concerned laymen as well as other ecologists around the world. A measure of the institute's productivity is the number of its studies accepted for publication—well over 300 scientific papers, theses and dissertations so far.

No ivory-tower types, IES scientists believe that their findings should be used to predict the consequences of human interactions with the environment—and that they should also help shape public policy. They serve as frequent advisers to New York's governor and its Department of Environmental Conservation on acid rain and other matters, and they keep up a running dialogue with key politicians and with officials of the federal Environmental Protection Agency. ("Some scientists don't give a damn about public policy," says Likens. "We do.")

As a result of more recent studies at Hubbard Brook, Likens and his collaborators proposed that no patch of forest land in the Northeast be clear-cut more than once every 75 years in order to give it time for proper recovery, a recommendation that the U.S. Forest Service has accepted in its planning. In the light of such contributions, perhaps it is no coincidence that Gene Likens was awarded the 1990 Distinguished Service Award of the American Institute of Biological Sciences, with a citation for "service to biological knowledge and its applications to the public well-being."

On a given day at IES, scientists are working on a variety of projects that range from their own backyard to locations around the country and overseas. By "bagging" tree branches at the Cary Arboretum, plant ecologist Gary Lovett and others analyze ways that ozone affects the ability of trees to obtain needed nutrients—a probable cause of forest decline in North America as well as Europe. Chemical ecologist Clive

Aquatic ecologists David Strayer and Stuart Findlay record data from the East branch of Wappinger Creek in Millbrook, site of long-term IES studies. (Ted Spiegel)

Jones studies the population dynamics of the gypsy moth in order to predict outbreaks of devastating defoliation. In another novel project, he has shown how lichen-eating snails in Israel's Negev Desert turn rocks into soil at a rate of three tons per acre each year. Aquatic ecologist David Strayer observes microorganisms as indicators of deteriorating groundwater quality. Wildlife ecologist Jay McAninch and others have studied ways of discouraging deer from browsing on cultivated plants, as well as their role in the spread of tick-borne Lyme disease.

IES ecologists increasingly collaborate with their counterparts from other countries, some of whom spend time at Cary as visiting scientists. IES also hosts a biennial conference at the arboretum, which in 1989 brought together 65 ecologists from the United States and abroad. Through one of its visiting associates, Juan Armesto of the University of Chile, the institute organized an international workshop in Santiago in 1990, at which 50 scientists from North and South America exchanged detailed findings on ecosystems in their respective temperate zones. Participants have hailed such conferences as models for future international collaboration on environmental problems, which keep mounting in a shrinking world.

Of IES projects under way, one strikes especially close to home: the study and management of NYBG's priceless 40-acre forest in the Bronx.

As early as 1908, Director Britton and other Garden officials were concerned enough about the "people problem" to put iron fencing along the forest's trails. By the 1920s they noticed with alarm that the hemlocks of their prized Hemlock Grove were failing to reproduce, and tried planting new ones without much long-term success.

After intervening decades of neglect, in 1984 IES was offered the challenge of managing the forest before time ran out. In analyzing the situation, Mark McDonnell, the project's leader, found that what had once been a mature woodland dominated by majestic hemlocks and oaks, some of them 250 to 300 years old, was gradually becoming a transitional forest that included tougher, younger, more opportunistic species like maple, cherry and birch, reaching for their own places in the sun.

"For years the trees had been under a tremendous amount of stress," says McDonnell. "People were walking off the trails, compacting the soil, disturbing the roots, carving their initials in trunks, setting fires. Some homeless individuals were actually living in the forest in rough shelters; a survivalist had excavated a series of tunnels with caches of food and weapons to retreat to, in the event of nuclear war, I guess. As recently as four years ago, platoons of ROTC students from Fordham University across the road climbed the Garden fence in full battle gear to use the woods for training maneuvers, and that certainly didn't help. We politely told the colonel he could hold his maneuvers somewhere else."

Adding to direct stresses from humans were indirect ones in the form of urbanization and air pollution. An accumulation of wax and oils on top of the soil—a suspected product of urban grime—kept the roots from getting sufficient water much of the year, and the resulting runoff increased erosion as well. An analysis of the soil itself showed some of the highest concentrations of heavy metals in the Northeast, deposited from industrial smokestacks and the leaded gas long burned by cars. The hemlocks in particular suffered, because their evergreen needles were exposed to airborne pollutants like ozone and sulfur dioxide all year long. They were also under attack from natural factors like scale insects, and in recent years the woolly adelgid, a new and even more ominous threat.

McDonnell and his colleagues drew up a management plan. They couldn't do much directly about air pollution, and they determined that spraying the forest with pesticides was impractical and could actually do more harm than good. Earlier Garden officials had proposed fencing off the entire forest, limiting its use and educational value, but the IES researchers decided to take a different tack: maintain only a few essential trails, restore the protective fences along them, make the place more attractive with benches, overlooks and educational signs, allow unwanted trails to revert to forest. At the same time they introduced regular patrolling, not so much by the Garden's uniformed security guards but by young forest employees wearing T-shirts and NYBG caps. These informal greeters were encouraged to say hello to visitors and offer to answer questions. They were also instructed to keep an eye on potential vandals, and to quickly repair or erase any evidences of damage or graffiti.

"It's worked," says McDonnell. "The message is clear that this is not a place to 'hang out,' but a piece of nature that should be respected. It's unbelievable. We've had practically no vandalism, and people enjoy the forest more. Sometimes they even repair a sign that someone else has knocked down, because they feel it's *their* forest."

◀ *The NYBG Forest, centerpiece of the Garden's grounds in the Bronx, is the subject of ongoing scientific and management studies.* (Allen Rokach/NYBG)

McDonnell thinks there may be a message in this for others who are trying to preserve small natural areas in the middle of cities.

Meanwhile, other educational roles for IES have blossomed at the arboretum itself, where scientists and teachers believe in getting the lessons of ecology out to the public at large. Behind Gifford House, the visitor center, a perennial garden of some 800 low-maintenance species and cultivars demonstrates that gardeners can enjoy both fine horticulture and sound ecology. An outdoor science center features an innovative "air pollution garden" where visiting adults and children can compare the effects of ozone on resistant and nonresistant plants. Near this are the "acid rain study ponds," 16 100-gallon tanks that demonstrate how varying amounts of acid affect plants and small creatures collected from local ponds. The outdoor science center has proved a popular part of the arboretum's educational program, which now includes an "eco-inquiry" curriculum for fifth and sixth graders (see box).

A feature of the Institute's Outdoor Science Center is a "walk-through pond ecosystem," a full-scale model in which students learn about food webs in aquatic environments. (Michael Doolittle)

Offerings for adults at Millbrook include certificate programs in landscape design and gardening as well as individual courses in ecology, botany, botanical illustration and nature photography. IES also holds special workshops in practical ecology and land management for landowners, landscape designers and other professionals. Particularly popular among residents of the region are Sunday afternoon ecological "walks and talks," and Friday seminars by visiting scientists, both open to the public at no charge.

Equally ambitious in its mission is the Institute of Economic Botany, which investigates new, little-known or underutilized plants for their usefulness in nutrition,

Students visiting the Institute of Ecosystem Studies observe the effects of simulated acid rain on organisms in 16 tanks maintained at different pH levels. (Alan Berkowitz)

(Ann Marie Gaynor)

Not-So-Mad Scientists

If children are asked to close their eyes and conjure up a "scientist," many come up with a predictable picture: a man with wild hair and even wilder eyes, dressed in a white lab coat, holding up a frothing test tube.

When the Institute of Ecosystem Studies was formulating its educational goals in the fall of 1985, it sought ways to teach scientific literacy through the process of inquiry—and, in so doing, to debunk the "mad scientist" stereotype. Working with IES ecologists, Kass Hogan, the institute's program leader in ecology education, developed a curriculum for fifth and sixth graders. Named "Eco-Inquiry," it encourages students to understand processes in nature and to explore their own capacities for finding original answers to research questions. At the same time, the curriculum works to develop intellectual curiosity, openmindedness, persistence and objectivity—attributes with long-term beneifts in both academic and everyday life.

Results were apparent from the start. Remarked one sixth grader: "I never thought scientists were like real people. I thought of them [as] I saw them on TV and books." Commented another: "I [liked best] learning how to think and act like a scientist when they find a problem or something puzzling."

Lessons often incorporate role-playing activities and games to make learning fun. In "Riddle-Me-This," the answers to riddles are suggestions for ingredients of a mini-ecosystem, a classroom terrarium:

I hold roots and earthworms, and other goodies, too. When you're standing outside, I even hold you. (Answer: soil.)

We're in colonies so tiny we can't usually be seen, but we sure can clean things up, if you know what I mean. (Answer: bacteria.)

Says Kass Hogan: "Eco-Inquiry helps kids understand ecological processes that support life on our planet. We hope that this knowledge of the world around them, coupled with insight into the roles of scientists in our society, will prepare young people to deal responsibly with the scientific and environmental issues of the coming decades."

energy, medicine and other fields. Founded in 1981 under the leadership of Ghillean Prance, then NYBG's senior vice-president for science, and now directed by Michael Balick, a leading economic botanist, the institute has turned the uncoordinated labors of a few specialists into a cohesive program of research and education, ranging from alternatives to the destruction of tropical rainforests to possible cures for cancer and AIDS.

Economic or "applied" botany, of course, is not a new idea—people have made practical use of plants ever since they first ate apples, threw branches on campfires and discovered that the leaves of certain plants were good for stomachaches. Nor were NYBG's founders unmindful of progress and possibilities in the field. Nathaniel Lord Britton himself experimented with a new plant source for rubber, though without success.

More fortunate was Alexander Anderson, whom Britton hired in 1901. While studying cereal grains, his specialty, he devised a method of heating them in sealed tubes. When he smashed a tube one day, the grains exploded—making Anderson the inventor of puffed wheat and puffed rice and, not long after, an honored beneficiary of Quaker Oats. In the mid-1920s, NYBG's Arlow Stout crossed native and Mediterranean species to produce new varieties of seedless grapes; he also doubled the amount of harvestable timber from poplars by developing a new strain that grew faster and proved more resistant to insects and disease. Of even greater consequence was the work of Bernard Dodge, a plant pathologist, who in 1928 began a series of investigations into a genus of fungus named *Neurospora*. In demonstrating its use as a tool for biochemical and genetic research, he paved the way for the Nobel-prize-winning work of others in discovering DNA.

The first economic botanist formally associated with the Garden as such was Henry Hurd Rusby, Professor of Botany and Materia Medica at Columbia Unversity's College of Pharmacy. In 1898 Rusby, an old plant-collecting friend of Britton's, was named honorary curator of the Garden's economic collections, a title he held until his death in 1940.

Most memorable to early Garden visitors were the exhibits Rusby and his colleagues persuaded various companies to donate for display in the Museum Building's galleries, which are now largely preempted by needed working space. Enthusiastic crowds, it was reported, lined up to see such marvels as "eleven varieties of peanuts, contributed by Messrs. James Chieves & Co.," and "our most important exhibit in the line of beverages: 12 wines, contributed by the H. T. Dewar & Sons Co." Rusby regretted that the latter didn't include the actual grapes from which the wines were made. He made up the deficit, however, with "fifty species of garden vegetables," which were somehat unappetizingly pickled in large jars of alcohol.

Perhaps the most popular presentation among younger visitors, donated by the American Chicle Company, was of "specimens representing chewing gum and its sources." Rusby reported that other displays of chewable plant products, including "preparations of opium used for these purposes" were still to be secured—which is perhaps just as well.

Since those innocent days, the work of NYBG's scientists has become considerably more sophisticated and far-reaching: as a division of the Garden's broad program in systematic botany, the Institute of Economic Botany has mounted what may be the most extensive effort of its kind based in the United States. Of its principal areas of research, the most urgent in view of world starvation is its quest for new or improved sources of food.

Underlying that quest, Balick points out, is an unsettling fact: while there are hundreds of thousands of kinds of plants, many with undeveloped possibilities, most of the world's food today comes from only 13 crop species like rice, wheat and

Digitalis, or foxglove, shown flowering on a South American mountainside, is the original source of the heart stimulant, and one of many plants from which modern medicines have been derived. (James Luteyn)

corn—and many of these, planted in large areas of monoculture, are vulnerable to epidemic disease and drought. Economic botanists seek other species that, through hybridizing, could contribute genetic resistance to such disasters. At the same time they try to broaden the crop base by finding and developing lesser-known plants that can serve as new, highly nutritional food sources in themselves.

One is *Myrciaria dubia*, a tall, scraggly shrub of the myrtle family that grows abundantly in the Peruvian Amazon. Known locally as "camu-camu," it has tart red fruits about the size of cherries that contain the highest content of vitamin C of any known species in the plant kingdom—30 times that of oranges. Juice from the berries makes a drink similar to lemonade, and the pulp can be used in ice cream, jellies and pastry. A popular local libation, "camu-camuchada," is prepared by mixing the juice and pulp with cane alcohol. Working with Peruvian scientists, IEB ecologist Charles Peters has published the results of three years of studies on camu-camu, including its potential for domestication and marketing on a broader scale.

Elsewhere in South America, IEB's David Williams has made germplasm collections of various relatives of peanuts to help plant breeders improve the peanut crop, on which many farmers depend. In the Andes, Stephen King has investigated native tubers that are staples of local diets, and has helped introduce these hardy, easily grown plants to other regions of the world. In the Far East, Tetsuo Koyama, the Garden's former director of Asiatic programs, has studied starch-producing species, including a hybrid canna lily that produces the largest known starch grains in the world.

As urgent as solving food shortages is the need for new sources of energy. Supplies of fossil fuels—coal, oil and natural gas, which had their origin in plants—are fast being used up. Moreover, the gases produced by burning them contain chemicals that are harmful to life, and that may contribute to long-term global warming through the greenhouse effect. In addition, most people throughout the developing world still use firewood to meet at least half of their energy needs, and countless others cut and burn woodlands to clear them for agriculture—contributing still more pollution to the atmosphere while destroying their forests at an alarming rate. A major challenge is to develop alternatives that will allow native populations to make a living from their forests without tearing them down.

In pursuing this challenge, Balick has become a leading expert on palms, which have long provided native peoples with both food and fuel, as well as a host of other products useful to man. In these multipurpose plants he sees a tremendous potential for "agroforestry," a system of tree farming that could replace current exploitation of the tropics with self-sustaining economies yielding far greater riches over an indefinite period of time.

Among the palms that Balick has studied, with the help of the U.S. Agency for International Development, is the babassu palm, which contains a veritable cornucopia of products in its tall, stately trunk and fountaining fronds. Chief sources of

▶

The rainforest in flames. IEB scientists seek to replace such scenes with methods to support the forest and its native population. (Robert Perron)

(Michael Balick)

Tree of a Thousand Uses

WHILE STUDYING NAtive palms in northeastern Brazil, IEB director Mike Balick and his colleagues were absorbed in examining seeds one day. Suddenly they sensed that someone had joined them and turned around to see who it was.

"There was our Shavanté Indian guide, covered with woven leaves from a babassu palm—pointing his shotgun straight at us," Balick recalls. "He had transformed himself into a regular walking blind that blended in perfectly with the surrounding vegetation. 'I wanted to show you how we hunt game,' he said proudly, lowering his gun. We were amazed at his skill in 'stalking' us without attracting our attention at all."

On another occasion one of the botanists accidentally cut his hand. As he reached for a first-aid kit, an Indian stepped forward and offered to handle the injury. "He stripped a leaf from a nearby babassu palm," says Balick, "scraped out some fluffy white pith and squeezed it on the wound. In a few seconds the bleeding stopped."

NYBG scientists soon learned that the babassu is a useful plant in many other respects. The Indians crack open the fruit to obtain a rich oil, which they use to cook food over fires made from the husks. The fruit, like that of larger coconuts, also provides tasty meat as well as milk, which they drink as a beverage or use in stewing meat and fish. Babassu oil lights the lamps in their simple homes, which are built of babassu wood, thatched with babassu fronds and equipped with babassu-fiber mats, baskets, sieves, fans and ornaments. Pith from the leaf stems is used as an antiseptic and styptic. The inner husk provides both flour and a home remedy for digestive complaints. Not even leftovers are wasted, but are pressed into use as animal feed and fishing bait.

In addition to its commercially valuable oil and charcoal, IEB scientists see in the babassu a promising source of alcohol, acetic acid, flour and fertilizer that could provide tropical regions with both sustenance and trade. The day might even come when a North American shopper will be able to buy products of the mighty babassu on her supermarket shelves.

these riches are its huge flower clusters, which resemble those of the coconut palm but, unlike that species, produce 200 or more smaller brown fruits. The hard husks contain kernels that yield a valuable oil which can be used for cooking and lighting, as a fuel for diesel engines and in the manufacture of cosmetics and soap. The babassu is also a source for many other products (see box). Moreover, it can grow on poor soil and thus holds much promise in reclaiming land ruined by deforestation. IEB scientists have combed various tropical countries to find superior strains, including a babassu species that yields 10 times more fruit and oil than the species commonly harvested, and another whose thinner husks are easier to crack in order to extract the oil-rich kernels.

Under IEB guidance, various babassu strains are hybridized to combine their best attributes; they are also tested with techniques to hasten germination and growth, including inoculating them with fungi that enable the roots to take up nutrients more efficiently from impoverished tropical soils. Plantations have been established in wasteland areas to examine the babassu's potential for reforestation.

The value of a tropical forest, like forests elsewhere, has been traditionally measured in two ways: how much the wood is worth on the market when cut, and how the cleared land can be converted to more "profitable" uses like ranching and agriculture—activities which, in reality, soon peter out because the poor soils underlying tropical forests are incapable of supporting them for more than a few years, forcing the farmers to move on.

Scientists from IEB and other organizations are finding strong evidence that species-rich forests can be much more valuable if left in place—that they can be both productive and protected at the same time. Rather than wholesale deforestation, which could completely denude the tropics in a matter of two or three decades, they propose a concept of conservation called "extractive reserves." Inspired by the example of small-scale rubber tappers, local inhabitants are encouraged to harvest renewable crops like Brazil nuts and other fruits, and at the same time are motivated by self-interest to protect the forest as the source of their livelihood.

An area of Brazilian forest after felling. When the slash has dried out, it will be burned to make way for short-term farming or cattle raising. (Robert Perron)

IEB anthropologist Christine Padoch discusses crops with a ribereño *farmer in Peru.* (Christine Padoch)

The concept has been dramatically illustrated by IEB's Charles Peters, who carried out studies in the small village of Mishana, not far from the city of Iquitos in Peru. Peters calculated that if all the salable timber in the area were cut at once, totally destroying the forest, it would bring a one-time profit to the villagers of only about $1,000 per hectare (roughly 2 1/2 acres) on delivery at the mill. In contrast, if the trees were left standing to produce 10 kinds of marketable fruits as well as rubber, the same hectare would net $422 a year indefinitely, and if selected trees were culled for timber every 20 years an additional dividend of $310 could be realized.

In other villages in the Peruvian Amazon, IEB anthropologist Christine Padoch and ecologist Wil de Jong have studied the techniques of the native *ribereños*, or river people, who, unlike conventional one-crop farmers, manage to keep plots of land productive with an ingenious diversity of crops over a cycle that may last 25 to 50 years.

First they clear small areas of forest, converting the larger trees to charcoal for sale in local markets, then burn the rest of the slash. In the nutrient-rich ashes around the blackened stumps they plant basic starch crops like manioc and plantain as well as corn, tomatoes and other vegetables for immediate use. At the same time they intersperse these with other crops like pineapples and cashew nuts, which will produce food for consumption and sale over the next three or four years, as well as seedlings of peach palm, umari, Brazil nut and other native trees that will start to take over later and continue to yield fruit for decades beyond that.

Thus the *ribereños'* multicrop gardens gradually turn into diverse orchards, whose products are sold as a principal source of income. As neighboring tree species move in as a result of natural succession, the orchards eventually revert to forest, when the cycle can be started all over again.

In these and other projects, IEB scientists learn much about plants by examining how native cultures use them, a field known as ethnobotany. It is an area that has provided particularly fruitful insights into medicine, whose therapies have been based on plants for thousands of years.

As early as 5,000 B.C., Chinese healers had developed extensive floral pharmacopoeias with which to treat patients. Such herbal "drugstores" later became commonplace in the Near East and medieval Europe, and for a long time the world depended largely on herbs for its medicinal needs. Nearly half of all modern pharmaceuticals now in use stem from botanical roots—aspirin from the willow, digitalis from foxglove, morphine from the opium poppy, quinine from cinchona, the anti-leukemia drug vincristine from the Madagascar periwinkle, the tranquilizer aconitine from monkshood, and the first oral contraceptive, norethindrone, from a wild Mexican yam.

While most drugs bought at a pharmacy today are the products of laboratory synthesis, the search for new cures and more effective treatments still depends on the complex molecular structures that plants have evolved. And the best hunting ground for these is the tropics, where a vast variety of species exist, and where many kinds of plants have long been used by native peoples to treat common ills.

Most of the drugs derived from plants in the past have come from tropical species, not only because the tropics offer greater diversity but because they lack the world's greatest insecticide—winter. "With so many insects around all the time trying to eat them, plants have had to evolve different chemical compounds to keep their natural predators at bay," says Scott Mori, a veteran of South American expeditions.. "Those are the active kinds of compounds that chemists are most interested in. All of them are toxic in one way or another; if they are to be used on humans successfully, it's mainly a question of dosage and control."

A classic example is curare, produced by a number of tropical vines to discourage predators, and found by native hunters to be an effective poison with which to tip their arrows in order to kill or stun their prey. Because it is a muscle relaxant, curare has proved invaluable in modern surgery and in treating muscle spasms. Another example is rotenone, commonly used as an insecticide, which originally came from a plant that South American Indians use to stun and catch fish.

An even more dramatic demonstration of the power of tropical plants is coca, a scraggly shrub that Amazonian and Andean natives have long used for food and medicinal purposes, often chewing its leaves to suppress both hunger and fatigue. Its nutritional value, including a high count of vitamins and minerals, represents an important contribution to the poor diet of many Andean tribes. The beneficial aspects of coca, however, are far overshadowed by the fact that one of the 50 or more alkaloids it contains is cocaine, which in concentrated form has become the basis of South America's—and the world's—most notorious industry. (Opium, heroin, marijuana and other mood-altering drugs, not to mention alcohol, are also derived from plants.)

In searching for sources of new chemicals, botanists pay special attention to species with such "biodynamic" qualities. Over generations, by trial and error, indigenous peoples have discovered how to use many of those chemicals for healing, hunting and other practical ends. And if a plant is bioactive in one way, scientists have found, the chances are several times greater that it may prove valuable to modern science in another way, possibly resulting in a medical breakthrough of some kind.

Much of IEB's effort in finding useful plants has involved working with local anthropologists and botanists, and, in particular, drawing on the knowledge of native shamans or medicine men. Because the wisdom of such healers has never been written down, because many are elderly and frail, and because there are few younger people willing to carry on their traditions, Balick and his colleagues must work quickly to obtain this knowledge before it is too late.

In pursuit of this goal, botanist Brian Boom spent five months in a tribal village of Chácobo Indians in Bolivia, collecting plants and interviewing residents. Of 360 spe-

A colorful outdoor market in Iquitos, Peru, offers a variety of fruits and other products harvested from the forest by native entrepreneurs. (Allen Rokach)

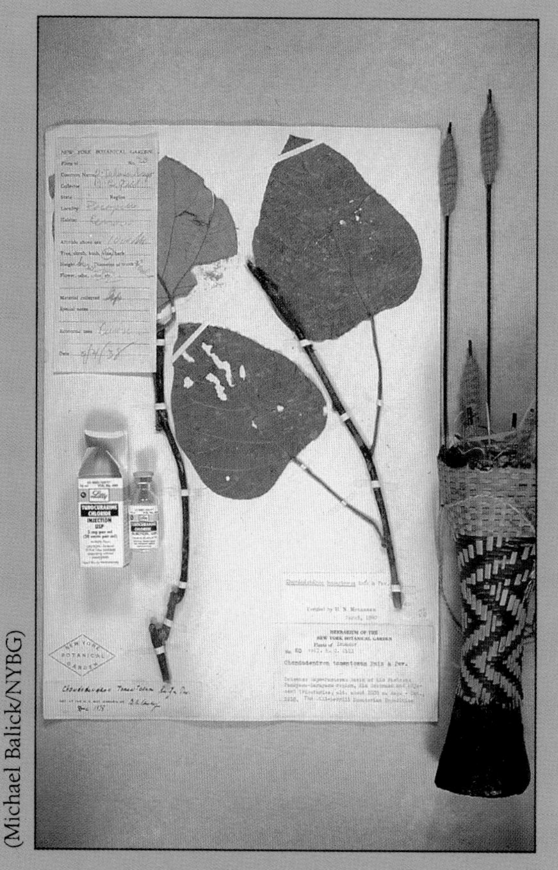

(Michael Balick/NYBG)

Prescriptions from Poisons

TOXIC CHEMICALS manufactured by plants can be lethal to man and other animals, but they can also turn out to be highly beneficial as drugs. A classic example is the story of curare, the "flying death," in which Garden scientists played a key role.

For almost four centuries, travelers to tropical America had brought back reports about native hunters who used poison-tipped darts and arrows to paralyze or kill their prey (as in the beautifully stylized painting of a bird hunter and his blowgun on an early Mexican vase, below). Intrigued by the potency of the mysterious substance, which tribesmen distilled from the bark of various vines, scientists tried using it to treat everything from rabies to epilepsy, with varying degrees of success.

By the early 1930s, Boris Alexander Krukoff, a botanist associated with the Garden, had become an expert in arrow poisons as a result of expeditions to Brazil sponsored by Merck & Company. Though he and others had identified promising components for drugs, confirmation was to come from an unexpected source.

In 1938 Richard Gill, an American businessman and amateur plant hunter, returned from an expedition to Ecuador with 26 different kinds of vines used in making curare. Gill sold his bulk samples to E. R. Squibb & Sons, who, after conducting chemical analyses, determined that the most potent poison came from a specimen labeled No. 20. Gill then submitted his specimens for analysis at NYBG, where Krukoff and his colleagues identified the source of the most powerful curare as a species of liana named *Chondrodendron tomentosum*.

An extract from the plant, d-tubocurarine, was developed into a neuromuscular blocking agent, which proved a boon in major surgery, spastic disorders, and other situations where it is vital to relax a patient's muscles temporarily. The historic specimen that helped transform an ancient killer into a modern life-saver is still in NYBG's herbarium, labeled "Richard Gill No. 20."

cies growing in the area, Boom discovered, the natives used no less than 305—85 percent—for food, fuel, construction and other purposes. Of the total, 174, or more than half, were employed for medicines, usually prepared by boiling the plant in water, pouring off the decoction and drinking it when it had cooled. Some 30 species were employed to relieve stomachaches, 26 to treat rheumatism, and lesser numbers for diarrhea, fever, skin infections, insect bites, snake bites, headaches, toothaches, head colds, hepatitis and appendicitis. A potion made from the bark of a small tree was drunk by women as a contraceptive. Other species were given to frail children to make them stronger, or to hyperactive children to calm them down.

Plants like these might solve major riddles of modern medicine, too. Since 1986, funded by a five-year contract with the National Cancer Institute with additional funds from the Metopolitan Life Foundation and the U.S. Agency for International Development, Balick, Douglas Daly and other colleagues have been searching the tropics for plants that may help in the treatment of cancer as well as the mounting epidemic of AIDS. The quest, which has taken them to a dozen countries in Latin America, is aimed at collecting 7,500 samples of as many as 4,000 selected species growing in various habitats. As collections are made, the specimens are shipped back to NCI's laboratory in Frederick, Maryland, where they are tested against more than a hundred strains of living cancers, as well as the AIDS virus, to see if they can act selectively to control diseased cells while sparing healthy ones.

Despite international pressures to stop destroying the South American rainforest, its future ultimately lies in the hands of the Amazonian nations, whose scientists are often impeded by insufficient funds, inadequate training and lack of government support. To help rectify the situation, IEB has secured grants to set up courses in cooperation with local institutions to teach young Brazilian botanists techniques for proper classification, conservation and management of their natural heritage.

Brian Boom, a specialist in native ethnobotany, on a research trip to the Venezuelan highlands. (Courtesy of Brian Boom)

In studying the rainforests, NYBG scientists are increasingly astounded at what highly diverse, and highly interdependent, assemblages of plants and animals they represent. Researchers are also impressed at how dependent these ecosystems are on the conditions in which they gradually evolved—and how vulnerable they may be to climatic changes brought on by man, which could unravel the fabric with diastrous results. Large areas of the Amazon basin and other tropical forests will have to be set aside in parks and preserves, and nondestructive forms of agroforestry encouraged, if their delicate ecosystems and great diversity are to survive.

Meanwhile, the vital process of identifying and evaluating the components of this diversity is moving slowly, due to lack of qualified scientists and funds. To date, only about 50 monographs have been published in *Flora Neotropica*, the most authoritative journal on the classification of New World tropical plants.

Says curator Scott Mori, who also acts as the director of the Organization for Flora Neotropica from its base at NYBG: "At the current rate of production it will take us another 380 years to complete the inventory, and long before that much of the rainforest and many of its species could be gone."

"Our use of species from the Amazon represents only the tip of a huge iceberg of possibilities," concludes IEB's Balick. "Only through accelerated programs, with increased support from international research and development agencies, will our ever-growing global family be able to benefit widely from this plethora of plants.

"Indeed, as we are now coming to realize, the future is in the forest."

The Amazonian rainforest, laced by silt-laden rivers, is a major focus of NYBG scientists—and disappearing at an alarming rate. (Robert Perron)

A New Century

"We have to give people what they can't get anywhere else."

"THE NEW YORK BOTANICAL GARDEN IS AT A HISTORICAL JUNCTURE, AND IT doesn't have anything to do with the fact that we're a hundred years old," says NYBG's president Gregory Long, looking out his office window at a Garden greening up with spring.

"We have a chance to take a great institution and turn it into one for the twenty-first century, to make it a stronger force in the world. As far as education is concerned, I don't think any institution has a greater opportunity. Do you realize that 20 million people live within a hundred miles of our gate?"

Unlike many who have charted the Garden's directions in the past, Long is not a scientist. Educated as an art historian at New York and Columbia universities, he worked for the Brooklyn Museum and the Metropolitan Museum of Art, then for 16 years successively headed the development efforts of three other respected institutions—the American Museum of Natural History, the New York Zoological Society and, most recently, the New York Public Library, where he is credited with much of the success of a $300 million capital campaign.

Since he took over NYBG's presidency in the spring of 1989, at the age of 42, Long has been trying to formulate images of the Garden's future, of the audiences that it must address, and of ways that funds can be found to reach those goals. He has brought in new vice-presidents for development and external affairs, created offices of planning and capital projects, given the Garden an initial facelifting to make it more appealing and informative to visitors. For the longer pull, his administration, under the direction of the board of managers, has held an ongoing series of staff

◀ *Floral banners and an outdoor café welcome visitors to the Garden's Museum Building.* (Marcia Stevens/NYBG)

A busy day at the Rockefeller Rose Garden. (Marcia Stevens/NYBG)

conferences aimed at identifying programs and improvements that will be needed in the decades ahead.

One of Long's first observations on his arrival was that the Garden's three major divisions—horticulture, science and education—were unquestionably leaders in their fields, but that they also seemed to be functioning as almost separate camps, not always comparing notes to see how their interests might mesh. A major priority became reaching a consensus on how they might work together.

"Of course we must continue to expand our classical role as a botanical garden, a museum of plants for study and public display," he says. "At the same time we must deal more effectively with the world in which we live. We have to devise a broader program in environmental education, one that makes use of all our departments and facilities—our scientists, our teachers, our collections, our gardens, our Conservatory, all of our 250 acres in the Bronx."

To Long and his colleagues, the key to the Garden's future is the same as its key to the past—research in systematic botany and ecology, old sciences that have taken on new importance as tools for dealing with today's pressures on the natural world. Without the hard data on species and habitats that they can provide, most scientists agree, conservation would be all but impossible.

A major part of the crisis is the fact that the world doesn't have enough systematists. It has been estimated that only about 150, at the most 200, professionals are truly competent in the study of plants in the tropics, where much of the earth's biodiversity lies (more than 20 of the total are at NYBG). To expand this cadre, the

▶

Lunch at the Tulip Tree Café. (Marcia Stevens/NYBG)

GARDEN SHOP
HERBARIUM

Garden intends to continue training its share of new systematic botanists, both in the Bronx and through cooperative programs in Latin America and other regions where the shortage is especially acute.

An equal challenge for NYBG will be explaining to ordinary people what systematic botany is, what it does, and why the study and preservation of species really matters.

"We have to try to figure out how to tell this story, both inside and outside the Garden's walls. We have to give people what they can't get anywhere else," says Long. "We must do a lot more active teaching with the plants in our Conservatory, which come from tropical and subtropical areas in which our scientists are experts. People are also concerned about the flora of North America and what's happening to it—acid rain and so forth. We have experts in those matters, too, and we have a priceless old forest right here on our grounds, which is a living example of many of the issues. It would be logical for us to be a center of information about the state of the North American forest."

NYBG president Gregory Long. (Marcia Stevens/NYBG)

Adds John Reed, NYBG's vice-president for education: "I don't know of any institution that does a thorough job of interpreting the relationships between plants and people, at least not yet. We have to convince people that without plants we wouldn't be here; we have to show them how plants work and what they do for us in our everyday lives, biologically, economically and esthetically."

To that end, Reed and others hope to restore the Garden's museum to its former prominence, among other things refurbishing it with a new gallery of ethnobotany to show how different cultures use plants, and another gallery that deals with the beauty of plants as seen in paintings, photographs, prints and books. In the museum, and elsewhere, they would like to see visitors watching videotapes of various plant processes, including the ingenious methods employed by carnivorous plants to capture and digest their prey, as well as demonstrations of gardening, ethnic foods and techniques that NYBG's plant hunters use in the field.

Like other institutions in a competitive, media-oriented world, the Garden must devise dramatic ways to get its message across. At the same time, its classic buildings are showing the effects of age. Plans have been drawn up to rehabilitate the Conservatory; to renovate the Museum Building and return parts of its interior to their original role as exhibit halls; to add a new wing to house the precious herbarium and library collections, which have outgrown their quarters and are endangered by lack of climate control.

Lastly, NYBG seeks fresh ways to interweave science and education with horticulture—to keep on improving and adding to its gardens, to introduce worthy new species and cultivars to the public, to make a visit to the Garden one that people will find rewarding not once but over and over again.

After all, without those beautiful flowers, a Sunday afternoon at The New York Botanical Garden wouldn't be much fun.

Planting seeds for tomorrow. (Allen Rokach/NYBG)

Appendix

The New York Botanical Garden

The New York Botanical Garden's grounds and buildings are owned by the city of New York, which supports their operation in part with public funds provided through the city's Department of Cultural Affairs. NYBG is a private, non-profit corporation that relies on individual, foundation and corporate gifts and grants, which in a typical year supply almost a third of its operating budget. Further help comes from the state of New York through the Natural Heritage Trust. The Garden's scientific and educational programs also receive grants from the National Science Foundation; the U.S. Agency for International Development; the Institute of Museum Services, a federal agency; the National Institutes of Health; the U.S. departments of State, Commerce and Education; the National Cancer Institute and other agencies and private foundations.

Chief Executive Officers of The New York Botanical Garden

Nathaniel Lord Britton	1896–1929
Elmer Drew Merrill	1930–1935
Marshall Avery Howe	1935–1936
William Jacob Robbins	1937–1957
William Campbell Steere	1958–1971
Howard Samuel Irwin	1971–1979
James McNaughton Hester	1980–1989
Gregory R. Long	1989–

Since the founding of the Garden, the post of chief executive officer has carried four different titles: Director in Chief (1896–1933), Director (1934-1968), Executive Director (1968–1972), President (1973 to date).

The Authors

Ogden Tanner is the author of *Gardening America, Garden Rooms* and a dozen more books on horticulture, nature, history and other subjects, as well as a contributor to *Smithsonian, Horticulture, Connoisseur* and other magazines. An architectural graduate of Princeton, he served as senior editor of *Architectural Forum* and Time-Life Books, where he edited the *Time-Life Encyclopedia of Gardening.* Tanner lives in New Canaan, Connecticut, and is a trustee of The Nature Conservancy's Connecticut Chapter.

Adele Auchincloss, a longtime NYBG volunteer and member of the Board of Managers, is a great-grandniece of the Garden's first president, Cornelius Vanderbilt II. A resident of Manhattan and Claryville, New York, she also serves as a trustee of the Natural Resources Defense Council. Her husband, Louis Auchincloss, is the distinguished novelist and chronicler of the New York scene.

The Photographer

The work of Allen Rokach, former NYBG staff photographer who now serves the Garden as a consultant, has appeared in many Garden publications as well as *Horticulture, National Geographic, Natural History* and other magazines. He is the co-author of *Focus on Flowers: Discovering and Photographing Beauty in Gardens and Wild Places* and founder and director of the Center for Nature Photography in Riverdale, New York.

Index